Euclidean and Non-Euclidean Geometries

EUCLIDEAN AND NON-EUCLIDEAN GEOMETRIES

DEVELOPMENT AND HISTORY

Marvin Jay Greenberg
University of California, Santa Cruz

W. H. FREEMAN AND COMPANY
San Francisco

Portraits by Henry Benson, 1973

Library of Congress Cataloging in Publication Data

Greenberg, Marvin J.
  Euclidean and non-Euclidean geometries.

  Bibliography: p. 291
    1. Geometry. 2. Geometry, Non-Euclidean.
3. Geometry—History.
4. Geometry, Non-Euclidean—History.
I. Title.
QA445.G84   516′.009   74-3491
ISBN 0-7167-0454-4

AMS subject classifications: 50-01, 50-03, 50A05, 50A10

Printed in the United States of America

10 9 8 7 6 5 4 3 2

To my son David Marcel Joseph

The moral of this book is:

Check your premises.

Contents

Preface

This book presents the discovery of non-Euclidean geometry and the subsequent reformulation of the foundations of Euclidean geometry as a suspense story. It has been written with several kinds of students in mind. Prospective high-school geometry teachers are presented with a rigorous treatment of the foundations of Euclidean geometry and an introduction to hyperbolic geometry (with emphasis on the Poincaré and Beltrami-Klein models). Liberal-arts students (particularly philosophy majors) are introduced to the history and philosophical implications of the discovery of non-Euclidean geometry. Mathematics majors are given challenging exercises and presented with an historical perspective that, sadly, is lacking in most mathematics texts.

I have used the development of non-Euclidean geometry to revive interest in the study of Euclidean geometry. I believe that this approach makes a traditional college course in Euclidean geometry more interesting. In order to identify the flaws in various attempted proofs of the Euclidean parallel postulate, we carefully examine the axiomatic foundations of Euclidean geometry; in order to prove the relative consistency of hyperbolic geometry, the properties of inversion in circles are studied; in order to justify János Bolyai's construction of the limiting parallel rays, cross ratios, harmonic tetrads, and perspectivities are introduced.

I have used modified versions of Hilbert's axioms for Euclidean geometry. Although some instructors and textbook writers prefer to bring in the real numbers at the beginning via ruler-and-protractor postulates, this requires a lengthy digression into the properties of the real number system. Hilbert's axioms, on the other hand, are close to the spirit of high-school geometry, and bright high-school students should be able to understand this book. Most of the subject is developed without reference to continuity; the few times that a continuity argument is required, an intuitive statement such as the circular continuity principle (p. 82), the elementary continuity principle (p. 82), or Archimedes' axiom suffices.

Dedekind's axiom is used here only in the discussion of hyperbolic geometry, to prove the existence of limiting parallel rays (of course, Dedekind's axiom is also needed to obtain an axiom system that is categorical).

A unique feature of this book is that new results are developed *in the exercises* and then built on in subsequent chapters. My experience teaching from earlier versions of this text convinced me that such a plan is crucial for the students' understanding (students not only learn by doing, they enjoy developing new results on their own). *If students do not do the exercises, they will have difficulty following subsequent chapters.* There are two sets of exercises in the first six chapters; the "major exercises" are the more difficult ones, which *all* students should attempt but which mathematics majors are more likely to solve. Hints are given for most of the exercises. The exercises in Chapter 7 present interesting supplementary topics, such as the classification of motions in Euclidean and hyperbolic planes. The "exercises" in Chapter 8 are unusual, consisting of philosophical and historical essay topics.

Terminology and notation are reasonably standard. I have followed W. Prenowitz and M. Jordan (*Basic Concepts of Geometry*) in using the term "neutral geometry" for the part of Euclidean geometry that is independent of the parallel postulate (the traditional name "absolute geometry" misleadingly implies that all other geometries depend on it). I have introduced the names "asymptotic" and "divergent" for the two types of parallels in hyperbolic geometry; I consider these a definite improvement over the welter of misleading names in the literature. The theorems, propositions, and figures are numbered by chapter; for example, Theorem 4.1 is the first theorem in Chapter 4. Such directives as "see De Leeuw" refer to the bibliography at the back of the book (the bibliography is arranged topically rather than strictly alphabetically).

I am happy to express appreciation to my students at the University of California, Santa Cruz, who struggled through earlier versions of this book, whose criticisms helped improve the manuscript, and whose enthusiasm boosted my morale; to Professor Don Chakerian for his suggestions after successfully testing my notes in his class at the University of California, Davis; to Professor István Fáry of the University of California, Berkeley, for contacting Hungarian mathematicians in an attempt to obtain a portrait of János Bolyai (apparently no authentic portrait exists); to Rosemary Stampfel for expertly typing much of the manuscript; to Professor Serge Lang of Yale University, for telling me that he wished he had studied hyperbolic geometry in college because he now needs it for his research on automorphic functions and number theory; to Rolf R. Loehrich for his provocative and supportive letters; and to all the friendly people at W. H. Freeman and Company who helped produce this book.

Passages from the correspondence of C. F. Gauss (to Taurinus and W. Bolyai), used throughout the book, are from Harold E. Wolfe's *Introduction to Non-Euclidean Geometry* (Holt, Rinehart and Winston, 1945).

March, 1974   Marvin Jay Greenberg

Euclidean and Non-Euclidean Geometries

Introduction

Let no one ignorant of geometry enter this door.
Entrance to Plato's Academy

Most people are unaware that around a century and a half ago a revolution took place in the field of geometry that was as scientifically profound as the Copernican revolution in astronomy and, in its impact, as philosophically important as the Darwinian theory of evolution. "The effect of the discovery of hyperbolic geometry on our ideas of truth and reality has been so profound," writes the great Canadian geometer H.S.M. Coxeter, "that we can hardly imagine how shocking the possibility of a geometry different from Euclid's must have seemed in 1820." Today, however, we have all heard of the space-time geometry in Einstein's theory of relativity. "In fact, the geometry of the space-time continuum is so closely related to the non-Euclidean geometries that some knowledge of [these geometries] is an essential prerequisite for a proper understanding of relativistic cosmology."

Euclidean geometry is the kind of geometry you learned in high school, the geometry most of us use to visualize the physical universe. It comes from the text by the Greek mathematician Euclid, the *Elements*, written around 300 B.C. Our picture of the physical universe based on this geometry was painted largely by Isaac Newton in the late seventeenth century.

Geometries that differ from Euclid's own arose out of a deeper study of *parallelism*. Consider this diagram of two rays perpendicular to segment PQ:

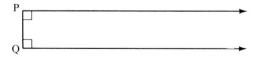

In Euclidean geometry the perpendicular distance between the rays remains equal to the distance from P to Q as we move to the right. However, in the early nineteenth century two alternative geometries were proposed. In hyperbolic geometry (from the Greek *hyperballein,* "to exceed") the distance between the rays increases. In elliptic geometry (from the Greek *elleipen,* "to fall short") the distance decreases and the rays eventually meet. These non-Euclidean geometries were later incorporated in a much more general geometry developed by C. F. Gauss and G. F. B. Riemann (it is this more general geometry that is used in Einstein's general theory of relativity).*

We will concentrate on Euclidean and hyperbolic geometries in this book. Hyperbolic geometry requires a change in only one of Euclid's axioms, and can be as easily grasped as high school geometry. Elliptic geometry, on the other hand, involves the new topological notion of "non-orientability," since all the points of the elliptic plane not on a given line lie on the same side of that line. This geometry cannot easily be approached in the spirit of Euclid, for it requires the prior development of projective geometry. I have thus relegated the discussion of elliptic geometry to a brief appendix. (Do not be misled by this, however; elliptic geometry is no less important than hyperbolic.) Riemannian geometry requires a thorough understanding of the differential and integral calculus, and is therefore beyond the scope of this book (it is discussed briefly in Appendix B).

Chapter 1 begins with a brief history of geometry in ancient times, and emphasizes the development of the axiomatic method by the Greeks. It presents Euclid's five postulates and concludes with Legendre's attempted proof of the fifth postulate. In order to detect the flaw in Legendre's argument (and in other arguments), it will be necessary to carefully reexamine the foundations of geometry. However, before we can do any geometry at all, we must be clear about some fundamental principles of logic. These are reviewed informally in Chapter 2. In this chapter we consider what constitutes a rigorous proof, giving special attention to the method of indirect proof, or *reductio ad absurdam.* Chapter 2 concludes with the very important notion of a *model* for an axiom system, illustrated by finite models for the axioms of incidence of points and lines in geometry.

Chapter 3 beings with a discussion of some flaws in Euclid's presentation of geometry. These are then repaired in a thorough presentation of

---

* Einstein's special theory of relativity, which is needed to study subatomic particles, is based on a simpler geometry of space-time due to H. Minkowski. The names "hyperbolic geometry" and "elliptic geometry" were coined by F. Klein; some authors misleadingly call these geometries "Lobachevskian" and "Riemannian," respectively.

David Hilbert's axioms (slightly modified) and their elementary consequences. You may become restless over the task of proving results that
appear self-evident. Nevertheless, this work is essential if you are to steer
safely through non-Euclidean space.

Our study of the consequences of Hilbert's axioms, with the exception
of the parallel postulate, is continued in Chapter 4; this study is called
neutral geometry. We will prove some familiar Euclidean theorems (such
as the exterior angle theorem) by methods different from those used by
Euclid, a change necessitated by gaps in Euclid's proofs. We will also
prove some theorems that Euclid would not recognize (such as the
Saccheri-Legendre theorem).

Supported by the solid foundation of the preceding chapters, we will
be prepared to analyze in Chapter 5 several important attempts to prove
the parallel postulate (in the exercises you will have the opportunity to
find flaws in still other attempts). Following that, your Euclidean con-
ditioning should be shaken enough so that in Chapter 6 we may explore
"a strange new universe," one in which triangles have the "wrong" angle
sums, rectangles do not exist, and parallel lines may diverge or converge
asymptotically. In doing so, we will see unfolding the historical drama
of the almost simultaneous discovery of hyperbolic geometry by Gauss,
J. Bolyai, and Lobachevsky in the early nineteenth century.

This geometry, however unfamiliar, is just as consistent as Euclid's.
This is demonstrated in Chapter 7 by studying three Euclidean models
that also aid in visualizing hyperbolic geometry. The Poincaré models
have the advantage that angles are measured in the Euclidean way; the
Beltrami-Klein model has the advantage that lines are represented by
segments of Euclidean lines. In Chapter 7 we will also discuss topics in
Euclidean geometry not usually covered in high school.

Finally, Chapter 8 takes up in a general way some of the philosophical
implications of non-Euclidean geometries. The presentation is deliberately
controversial, and the essay topics are intended to stimulate further
thought and reading.

It is very important that you do *all* the exercises, since new results
are developed in the exercises and then built on in subsequent chapters.
By working all the exercises, you may come to enjoy geometry as
much as I do.

Euclid's Geometry　1

The postulate on parallels ... was in antiquity the final solution of a problem that
must have preoccupied Greek mathematics for a long period before Euclid.
Hans Freudenthal

The Origins of Geometry

The word "geometry" comes from the Greek *geometrein* (*geo*: earth, and
metrein: to measure); geometry was originally the science of measuring
land. The Greek historian Herodotus (5th century B.C.) credits Egyptian
surveyors with having originated the subject of geometry, but other ancient
civilizations (Babylonian, Hindu, Chinese) also possessed much geometric
information.

Ancient geometry was actually a collection of rule-of-thumb pro-
cedures arrived at through experimentation, observation of analogies,
guessing, and occasional flashes of intuition. In short, it was an empirical
subject in which approximate answers were usually sufficient for practical
purposes. The Babylonians of 2,000 to 1,600 B.C. considered the circumference of a circle to be three times the diameter, i.e., they took π to
be equal to 3. This was the value given by the Roman architect Vitruvius
and it is found in the Chinese literature as well. It was even considered
sacred by the ancient Jews and sanctioned in scripture (I Kings 7:23)—an
attempt by Rabbi Nehemiah to change the value of π to $\frac{22}{7}$ was rejected.
The Egyptians of 1,800 B.C., according to the Rhind papyrus, had the
approximation $\pi \sim (\frac{16}{9})^2 \sim 3.1604.$*

Sometimes the Egyptians guessed correctly, other times not. They found
the correct formula for the volume of a frustrum of a square pyramid—
a remarkable accomplishment. On the other hand, they thought that a
formula for area that was correct for rectangles applied to *any* quadrilateral. Egyptian geometry was not a science in the Greek sense, only a

---

* In recent years π has been approximated to a very large number of decimal places by
computers; to five places, π is approximately 3.14159. In 1789 Johann Lambert proved that
π was not equal to any fraction (rational number), and in 1882 Lindemann proved that π is
a transcendental number, in the sense that it does not satisfy any algebraic equation with
rational coefficients.

grab bag of rules for calculation without any motivation or justification.

The Babylonians were much more advanced than the Egyptians in arithmetic and algebra. Moreover, they knew the Pythagorean theorem—in a right triangle the square of the length of the hypotenuse is equal to the sum of the squares of the lengths of the legs—long before Pythagoras was born. Recent research by Otto Neugebauer has revealed the heretofore unknown Babylonian algebraic influence on Greek mathematics.

However, the Greeks, beginning with Thales of Milete, insisted that geometric statements be established by deductive reasoning rather than by trial and error. Thales was familiar with the computations, partly right and partly wrong, handed down from Egyptian and Babylonian mathematics. In determining which results were correct, he developed the first logical geometry (Thales is also famous for having predicted the eclipse of the sun in 585 B.C.). The orderly development of theorems by *proof* was characteristic of Greek mathematics and entirely new.

The systematization begun by Thales was continued over the next two centuries by Pythagoras and his disciples. Pythagoras was regarded by his contemporaries as a religious prophet. He preached the immortality of the soul and reincarnation. He organized a brotherhood of believers that had its own purification and initiation rites, followed a vegetarian diet, and shared all property communally. The Pythagoreans differed from other religious sects in their belief that elevation of the soul and union with God are achieved by the study of music and mathematics. In music, Pythagoras calculated the correct ratios of the harmonic intervals. In mathematics, he taught the mysterious and wonderful properties of numbers. Book VII of Euclid's *Elements* is the text of the theory of numbers taught in the Pythagorean school.

The Pythagoreans were greatly shocked when they discovered irrational lengths, such as $\sqrt{2}$ (see Chapter 2, pp. 35–37). At first they tried to keep this discovery secret. The historian Proclus wrote: "It is well known that the man who first made public the theory of irrationals perished in a shipwreck, in order that the inexpressible and unimaginable should ever remain veiled." Since the Pythagoreans did not consider $\sqrt{2}$ a number, they transmuted their algebra into geometric form in order to represent $\sqrt{2}$ and other irrational lengths by segments ($\sqrt{2}$ by a diagonal of the unit square).

The systematic foundation of plane geometry by the Pythagorean school was brought to a conclusion around 400 B.C. in the *Elements* by the mathematician Hippocrates (not to be confused with the physician of the same name). Although this treatise has been lost, we can safely say that it covered most of Books I-IV of Euclid's *Elements,* which appeared about a century later. The Pythagoreans were never able to develop a theory of proportions that was also valid for irrational lengths. This was later achieved by Eudoxus, whose theory was incorporated into Book V of Euclid's *Elements.*

The fourth century B.C. saw the flourishing of Plato's Academy of science and philosophy (founded about 387 B.C.). In the *Republic* Plato wrote, "The study of mathematics develops and sets into operation a mental organism more valuable than a thousand eyes, because through it alone can truth be apprehended." Plato taught that the universe of ideas is more important than the material world of the senses, the latter being only a shadow of the former. The material world is an unlit cave on whose walls we see only shadows of the real, sunlit world outside. The errors of the senses must be corrected by concentrated thought, which is best learned by studying mathematics. The Socratic method of dialog is essentially that of indirect proof, by which an assertion is shown to be invalid if it leads to a contradiction. Plato repeatedly cited the proof for the irrationality of the length of a diagonal of the unit square as an illustration of the method of indirect proof (the *reductio ad absurdum*—see Chapter 2, pp. 34–37). The point is that this irrationality of length could never have been discovered by physical measurements, which always include a small experimental margin of error.

Euclid was a disciple of the Platonic school. Around 300 B.C. he produced the definitive treatment of Greek geometry and number theory in his thirteen-volume *Elements.* In compiling this masterpiece Euclid built on the experience and achievements of his predecessors in preceding centuries: on the Pythagoreans for Books I-IV, VII, and IX, Archytas for Book VIII, Eudoxus for Books V, VI, and XII, and Theaetetus for Books X and XIII. So completely did Euclid's work supersede earlier attempts at presenting geometry that few traces remain of these efforts. It's a pity that Euclid's heirs have not been able to collect royalties on his work, for he is the most widely read author in the history of mankind. His approach to geometry has dominated the teaching of the subject for over two thousand years. Moreover, the axiomatic method used by Euclid is the prototype for all of what we now call "pure mathematics." It is pure in the sense of "pure thought": no physical experiments need be performed to verify that the statements are correct—only the reasoning in the demonstrations need be checked.

Euclid's *Elements* is pure also in that the work includes no practical applications. Of course, Euclid's geometry has had an enormous number of applications to practical problems in engineering, but they are not mentioned in the *Elements*. According to legend, a beginning student of geometry asked Euclid, "What shall I get by learning these things?" Euclid called his slave, saying, "Give him a coin, since he must make gain out of what he learns." To this day, this attitude toward application persists among many pure mathematicians—they study mathematics for its own sake, for its intrinsic beauty and elegance.

Surprisingly enough, as we will see later, pure mathematics often turns out to have applications never dreamt of by its creators—the "impractical" outlook of pure mathematicians is ultimately useful to society. Moreover, those parts of mathematics that have not been "applied" are also valuable to society, either as aesthetic works comparable to music and art or as contributions to the expansion of man's consciousness and understanding.*

The Axiomatic Method

Mathematicians can make use of trial and error, computation of special cases, inspired guessing, or any other way to discover theorems. The axiomatic method is a method of *proving* that results are correct. Some of the most important results in mathematics were originally given only incomplete proofs (we shall see that even Euclid was guilty of this). No matter—correct proofs would be supplied later (sometimes much later) and the mathematical world would be satisfied.

So proofs give us assurance that results are correct. In many cases they also give us more *general* results. For example, the Egyptians and Hindus knew by experiment that if a triangle has sides of lengths 3, 4, and 5, it is a right triangle. But the Greeks proved that if a triangle has sides of lengths a, b and c, and if $a^2 + b^2 = c^2$, then the triangle is a right triangle. It would take an infinite number of experiments to check this result (and, besides, experiments only measure things approximately). Finally, proofs give us tremendous insight into relationships among different things we are studying, forcing us to organize our ideas in a coherent way.

---

* For more detailed information on ancient mathematics, see Bartel van der Waerden's *Science Awakening* (Oxford University Press, 1961).

What is the axiomatic method? If I wish to persuade you by *pure reasoning* to believe some statement S_1, I could show you how this statement follows logically from some other statement S_2 that you may already accept. However, if you don't believe S_2, I would have to show you how S_2 follows logically from some other statement S_3. I might have to repeat this procedure several times until I reach some statement that you already accept, one I do not need to justify. That statement plays the role of an *axiom* (or *postulate*). If I cannot reach a statement that you will accept as the basis of my argument, I will be caught in an "infinite regress," giving one demonstration after another without end.

So there are two requirements that must be met for us to agree that a proof is correct:

REQUIREMENT 1.   Acceptance of certain statements called "axioms," or "postulates," without further justification.

REQUIREMENT 2.   Agreement on how and when one statement "follows logically" from another, i.e., agreement on certain rules of reasoning.

Euclid's monumental achievement was to single out a few simple postulates, statements that were acceptable without further justification, and then to deduce from them 465 propositions, many complicated and not at all intuitively obvious, which contained all the geometric knowledge of his time. One reason the *Elements* is such a beautiful work is that so much has been deduced from so little.

Undefined Terms

We have been discussing what is required for us to agree that a proof is correct. Here is one requirement that we took for granted:

REQUIREMENT O.   Mutual understanding of the meaning of the words and symbols used in the discourse.

There should be no problem in reaching mutual understanding so long as we use terms familiar to both of us and use them consistently. If I use an unfamiliar term, you have the right to demand a *definition* of this term. Definitions cannot be given arbitrarily; they are subject to the rules of reasoning referred to (but not specified) in Requirement 2. If, for example, I define a right angle to be a 90° angle, and then define a 90° angle to be a right angle, I would violate the rule against *circular reasoning.*

Also, we cannot define every term that we use. In order to define one term we must use other terms, and to define these terms we must use still other terms, and so on. If we were not allowed to leave some terms *undefined,* we would get involved in infinite regress.

Euclid did attempt to define all geometric terms. He defined a "straight line" to be "that which lies evenly with the points on itself." This definition is not very useful; to understand it you must already have the image of a line. So it is better to take "line" as an undefined term. Similarly, Euclid defined a "point" as "that which has no part"—again, not very informative. So we will also accept "point" as an undefined term. Here are the five undefined geometric terms that are the basis for defining all other geometric terms in plane Euclidean geometry:

point
line
lie on (as in "two points *lie on* a unique line")
between (as in "point C is *between* points A and B")
congruent

For solid geometry, we would have to introduce a further undefined geometric term, "plane," and extend the relation "lie on" to allow points and lines to lie on planes. *In this book* (*unless otherwise stated*) *we will restrict our attention to plane geometry,* i.e., to one single plane. So we define *the plane* to be the set of all points and lines, all of which are said to "lie on" it.

There are expressions that are often used synonymously with "lie on." Instead of saying "point P *lies on* line *l*," we sometimes say "*l passes through* P" or "P is *incident* with *l*." If point P lies on both line *l* and line *m*, we say that "*l* and *m have* point P *in common*" or that "*l* and *m intersect* (or *meet*) in the point P."

The second undefined term, "line," is synonymous with "straight line." The adjective "straight" is confusing when it modifies the noun "line," so we won't use it. Nor will we talk about "curved lines." Although the word "line" will not be defined, its use will be restricted by the axioms for our geometry. For instance, one axiom states that two given points lie on only one line. Thus, in the figure below *l* and *m* could not represent lines in our geometry, since they both pass through the points P and Q. You would have to call *l* and *m* "curves," not "lines."

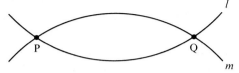

Figure 1.1

There are other mathematical terms that we will use that should be added to our list of undefined terms, since we won't define them; they have been omitted because they are not specifically geometric in nature, but are rather what Euclid called "common notions." Nevertheless since there may be some confusion about these terms, a few remarks are in order.

The word "set" is fundamental in all of mathematics today; it is now used in elementary schools, so undoubtedly you are familiar with its use. Think of it as a "collection of objects." Two related notions are "belonging to" a set or "being an element (or member) of" a set, as in our convention that all points and lines *belong* to the plane. If every element of a set S is also an element of a set T, we say that S is "contained in" or "part of" or "a subset of" T. For instance, the set of all children is a subset of the set of all people.

In the language of sets we say that sets S and T are *equal* if every member of S is a member of T, and vice-versa. For example, the set S of all authors of Euclid's *Elements* is (presumably) equal to the set whose only member is Euclid. Thus, "equal" means "identical."

Euclid used the word "equal" in a different sense, as in his assertion that "base angles of an isosceles traingle are *equal*." He meant that base angles of an isoceles triangle have an equal number of degrees, not that they are identical angles. So to avoid confusion we will not use the word "equal" in Euclid's sense. Instead, we will use the undefined term "congruent" and say that "base angles of an isosceles triangle are *congruent*." Similarly, we don't say that "if AB *equals* AC, then △ABC is isosceles." (If AB *equals* AC, following our use of the word "equals," △ABC is not a triangle at all, only a segment.) Instead, we would say that "if AB is *congruent* to AC, then △ABC is *isosceles*." This use of the undefined term "congruent" is more general than the one to which you are accustomed; it applies not only to triangles but to angles and segments as well. To understand the use of this word, picture congruent objects as "having the same size and shape."

Of course, we must specify (as Euclid did in his "common notions") that "a thing is congruent to itself," and that "things congruent to the same thing are congruent to each other." Statements like these will later be included among our axioms of congruence (Chapter 3).

The list of undefined geometric terms given above is due to David Hilbert (1862–1943). His treatise, *The Foundations of Geometry* (1899), not only clarified Euclid's definitions but also filled in the gaps in some of Euclid's proofs. Hilbert recognized that Euclid's proof for the side-angle-side criterion of congruence in triangles was based on an unstated assumption (the principle of superposition), and that this criterion had to be treated as an axiom. He also built on the earlier work of Moritz Pasch, who in 1882 published the first rigorous treatise on geometry; Pasch made explicit Euclid's unstated assumptions about betweenness. Some other mathematicians who worked to establish rigorous foundations for Euclidean geometry are: G. Peano, M. Pieri, G. Veronese, O. Veblen, G. de B. Robinson, E. V. Huntington, and H. G. Forder. These mathematicians used lists of undefined terms different from the one used by Hilbert. Pieri used only two undefined terms (as a result, however, his axioms were more complicated).

Euclid's First Four Postulates

Euclid based his geometry on five fundamental assumptions, called *axioms* or *postulates*.

EUCLID'S POSTULATE I.   For every point P and for every point Q not equal to P there exists a unique line *l* that passes through P and Q.

This postulate is sometimes expressed informally by saying "two points determine a unique line." We will denote the unique line that passes through P and Q by $\overleftrightarrow{PQ}$.

To state the second postulate, we must make our first definition.

DEFINITION.   Given two points A and B. The *segment* AB is the set whose members are the points A and B and all points that lie on the line $\overleftrightarrow{AB}$ and are between A and B. The two given points A and B are called the *endpoints* of the segment AB.*

* Warning on notation: In many high school geometry texts the notation $\overline{AB}$ is used for "segment AB."

Segment A B

Line $\overset{\longleftrightarrow}{A\,B}$

Figure 1.2

EUCLID'S POSTULATE II.   For every segment AB and for every segment CD there exists a unique point E such that B is between A and E and segment CD is congruent to segment BE.

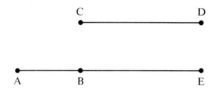

Figure 1.3
CD $\cong$ BE.

This postulate is sometimes expressed informally by saying that "any segment AB can be extended by a segment BE congruent to a given segment CD." Notice that in this postulate we have used the undefined term "congruent" in the new way, and we use the usual notation CD $\cong$ BE to express the fact that CD is congruent to BE.

In order to state the third postulate, we must introduce another definition.

DEFINITION.   Given two points O and A. The set of all points P such that segment OP is congruent to segment OA is called a *circle* with O as *center,* and each of the segments OP is called a *radius* of the circle.

It follows from our previously mentioned congruence axiom ("a thing is congruent to itself") that OA $\cong$ OA, so A is also a point on the circle just defined.

EUCLID'S POSTULATE III.   For every point O and every point A not equal to O there exists a circle with center O and radius OA.

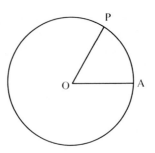

Figure 1.4

Circle with center O and radius OA.

Actually, because we are using the language of sets rather than that of Euclid, it is not really necessary to assume this postulate; it is a consequence of set theory that the set of all points P with OP $\cong$ OA exists. Euclid had in mind *drawing* the circle with center O and radius OA, and this postulate tells you that such a drawing is allowed, for example, with a compass. Similarly, in Postulate II you are allowed to extend segment AB by drawing segment BE with a straightedge. Our treatment "purifies" Euclid by eliminating all references to drawing.*

DEFINITION.   The *ray* $\overrightarrow{AB}$ is the following set of points lying on the line $\overleftrightarrow{AB}$: those points that belong to the segment AB and all points C such that B is between A and C. The ray $\overrightarrow{AB}$ is said to *emanate* from A and to be *part* of line $\overleftrightarrow{AB}$.

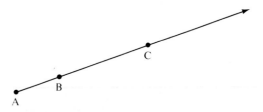

Figure 1.5

Ray $\overrightarrow{AB}$.

*However, it is, a fascinating mathematical problem to determine just what geometric constructions are possible using only a compass and straightedge. Not until the nineteenth century was it proved that such constructions as trisecting an arbitrary angle, squaring a circle, or doubling a cube were impossible using only a compass and straightedge. Pierre Wantzel proved this by translating the geometric problem into an algebraic problem; he showed that straightedge and compass constructions correspond to a solution of certain algebraic equations using only the operations of addition, subtraction, multiplication, division, and extraction of square roots. For the particular algebraic equations obtained from, say, the problem of trisecting an arbitrary angle, such a solution is impossible because cube roots are needed. Of course, it is possible to trisect angles using other instruments, such as a marked straightedge and compass (see Kutuzov, Moise, Eves, or the major exercises at the end of this chapter).

DEFINITION.   Rays $\overrightarrow{AB}$ and $\overrightarrow{AC}$ are *opposite* if they are distinct, if they emanate from the same point A, and if they are part of the same line $\overleftrightarrow{AB} = \overleftrightarrow{AC}$.

Figure 1.6
Opposite rays.

DEFINITION.   An "*angle* with *vertex* A" is a point A together with two nonopposite rays $\overrightarrow{AB}$ and $\overrightarrow{AC}$ (called the *sides* of the angle) emanating from A.*

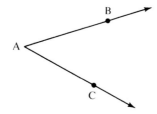

Figure 1.7
Angle with vertex A.

We use the notations ∢ A, ∢ BAC, or ∢ CAB for this angle.

DEFINITION.   If two angles ∢ BAD and ∢ CAD have a common side $\overrightarrow{AD}$ and the other two sides $\overrightarrow{AB}$ and $\overrightarrow{AC}$ form opposite rays, the angles are *supplements* of each other, or *supplementary angles.*

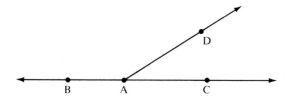

Figure 1.8
Supplementary angles.

---

* According to this definition, there is no such thing as a "straight angle." We eliminated this expression because most of the assertions we will make about angles do not apply to "straight angles."

DEFINITION.   An angle ⊀ BAD is a *right angle* if it has a supplementary angle to which it is congruent.

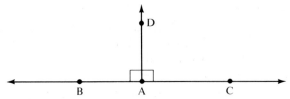

Figure 1.9
Right angles. ⊀ BAD ≅ ⊀ CAD.

We have thus succeeded in defining a right angle without referring to "degrees," by using the undefined notion of congruence of angles. "Degrees" will not be introduced formally until Chapter 4, although we will occasionally refer to them in informal discussions.

We can now state Euclid's fourth postulate.

EUCLID'S POSTULATE IV.   All right angles are congruent to each other.

This postulate expresses a sort of homogeneity; even though two right angles may be "very far away" from each other, they nevertheless "have the same size." The postulate therefore provides a natural standard of measurement for angles.*

The Parallel Postulate

Euclid's first four postulates have always been readily accepted by mathematicians. The fifth (parallel) postulate, however, was highly controversial until the nineteenth century. In fact, as we shall see later, consideration of alternatives to Euclid's parallel postulate resulted in the development of non-Euclidean geometries.

---

* On the contrary, there is no natural standard of measurement for *lengths* in Euclidean geometry. Units of length (one foot, one meter, etc.) must be chosen arbitrarily. The remarkable fact about hyperbolic geometry, on the other hand, is that it does admit a natural standard of length—see Chapter 6.

At this point we are not going to state the fifth postulate in its original form, as it appeared in the *Elements*. The reason is that we want to define our terms very carefully—if we are to avoid unnecessary difficulties. For greater clarity, we have already made some changes in Euclid's presentation, such as the change from calling segments (or angles) "equal" to calling them "congruent." Defining all the terms needed to make the original version of Euclid's fifth postulate intelligible would take us too far afield, and we will postpone stating it until Chapter 4.

Instead, we will present a simpler postulate (which we will later show is logically equivalent to Euclid's original). This version is sometimes called *Playfair's postulate* because it appeared in John Playfair's presentation of Euclidean geometry, published in 1795—although it was referred to much earlier by Proclus (410–485 A.D.). We will call it *the Euclidean parallel postulate* because it distinguishes Euclidean geometry from other geometries based on parallel postulates. The most important definition in this book is the following:

DEFINITION.   Two lines *l* and *m* are *parallel* if they do not intersect, i.e., if no point lies on both of them. We denote this by $l \parallel m$.

Notice first that we assume the lines lie in the same plane (because of our convention that all points and lines lie in one plane, unless stated otherwise). Notice secondly what the definition does *not* say: it does not say that the lines are equidistant, i.e., it does not say that the distance between the two lines is everywhere the same. Don't be misled by drawings of parallel lines in which the lines appear to be equidistant. We want to be rigorous here and so should not introduce into our proofs assumptions that have not been stated explicitly. At the same time, don't jump to the conclusion that parallel lines are *not* equidistant. We are not committing ourselves either way and shall reserve judgment until we study the matter further. At this point, the only thing we know for sure about parallel lines is that they do not meet.

THE EUCLIDEAN PARALLEL POSTULATE.   For every line *l* and for every point P that does not lie on *l* there exists a unique line *m* through P that is parallel to *l*.

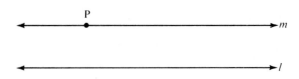

Figure 1.10

Lines *l* and *m* are parallel.

Why should this postulate be so controversial? It may seem "obvious" to you, perhaps because you have been conditioned to think in Euclidean terms. However, if we consider the axioms of geometry as abstractions from experience, we can see a difference between this postulate and the other four. The first two postulates are abstractions from our experiences drawing with a straightedge; the third postulate derives from our experiences drawing with a compass. The fourth postulate is perhaps less obvious as an abstraction, nevertheless it derives from our experiences measuring angles with a protractor (where the sum of supplementary angles is 180°, so that if supplementary angles are congruent to each other, they must each measure 90°).

The fifth postulate is different in that we cannot verify empirically whether two lines meet, since we can draw only segments, not lines. We can extend the segments further and further to see if they meet, but we cannot go on extending them forever. Our only recourse is to verify parallelism indirectly, by using criteria other than the definition.

What is another criterion for *l* to be parallel to *m*? Euclid suggested drawing a *transversal*, i.e., a line *t* that intersects both *l* and *m* in distinct points, and measuring the number of degrees in the interior angles α and β on one side of *t*. Euclid predicted that if the sum of angles α and β turns out to be less than 180°, the lines (if produced sufficiently far) would meet on the same side of *t* as angles α and β. This, in fact, is the content of Euclid's fifth postulate.

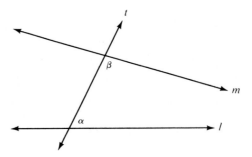

Figure 1.11

The trouble with this criterion for parallelism is that we are making an assumption that turns out to be logically equivalent to the parallel postulate stated above (see Equivalence of Parallel Postulates, Chapter 4). So we cannot use this criterion to convince ourselves of the correctness of the parallel postulate—that would be circular reasoning. Euclid himself recognized the questionable nature of the parallel postulate, for he postponed using it for as long as he could (until the proof of his 29th proposition).

Attempts to Prove the Parallel Postulate

Remember that an axiom was originally supposed to be so simple and intuitively obvious that no one could doubt its validity. From the very beginning, however, the parallel postulate was attacked as insufficiently plausible to qualify as an unproved assumption. For two thousand years mathematicians tried to derive it from the other four postulates or to replace it with another postulate, one more self-evident. All attempts to derive it from the first four postulates turned out to be unsuccessful because the so-called proofs always entailed a hidden assumption that was unjustifiable. The substitute postulates, purportedly more self-evident, turned out to be logically equivalent to the parallel postulate, so that nothing was gained by the substitution. We will examine these attempts in detail in Chapter 5, for they are very instructive. For the moment, let us consider one such effort.

The Frenchman Adrien Marie Legendre (1752–1833) was one of the best mathematicians of his time, contributing important discoveries to many different branches of mathematics. Yet he was so obsessed with proving the parallel postulate that over a period of 29 years he published one attempt after another in different editions of his *Éléments de Géométrie.**
Here is one attempt:

Given P not on line *l*. Drop perpendicular PQ from P to *l* at Q. Let *m* be the line through P perpendicular to $\overleftrightarrow{PQ}$. Then *m* is parallel to *l*, since *l* and *m* have the common perpendicular $\overleftrightarrow{PQ}$. Let *n* be any line through

---

* Davies' translation of the *Elements* was the most popular geometry textbook in the United States during the nineteenth century. Legendre is best-known for the method of least squares in statistics, the law of quadratic reciprocity in number theory, and the Legendre polynomials in differential equations. His attempts to prove the parallel postulate led to two important theorems in neutral geometry (see Chapter 4).

Adrien Marie Legendre

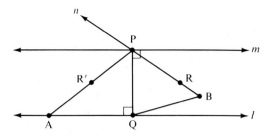

Figure 1.12

P distinct from m and $\overleftrightarrow{PQ}$. We must show than n meets l. Let $\overrightarrow{PR}$ be a ray of n between $\overrightarrow{PQ}$ and a ray of m emanating from P. There is a point R' on the opposite side of $\overrightarrow{PQ}$ from R such that $\measuredangle\, QPR' \cong \measuredangle\, QPR$. Then Q lies in the interior of $\measuredangle\, RPR'$. Since line l passes through the point Q interior to $\measuredangle\, RPR'$, l must intersect one of the sides of this angle. If l meets side $\overrightarrow{PR}$, then certainly l meets n. Suppose l meets side $\overrightarrow{PR'}$ at a point A. Let B be the unique point on side $\overrightarrow{PR}$ such that PA $\cong$ PB. Then $\triangle PQA = \triangle PQB$ (SAS); hence $\measuredangle\, PQB$ is a right angle, so that B lies on l (and n).

You may feel that this argument is plausible enough. Yet how could you tell if it is correct? You would have to justify each step, first defining each term carefully. For instance, you would have to define what was meant by two lines being "perpendicular"—otherwise, how could you justify the assertion that lines l and m are parallel simply because they have a common perpendicular? (You would first have to prove that as a separate theorem, if you could.) You would have to justify the side-angle-side (SAS) criterion of congruence in the last statement. You would have to define the "interior" of an angle, and prove that a line through the interior of an angle must intersect one of the sides. In proving all of these things, you would have to be sure to use only the first four postulates and not any statement equivalent to the fifth, otherwise the argument would be circular.

Thus there is a lot of work that must be done before we can detect the flaw. In the next few chapters we will do this preparatory work so that we can confidently decide whether or not Legendre's proposed proof is valid. (Legendre's argument contains several statements that cannot be proved from the first four postulates.) As a result of this work we will be better able to understand the foundations of Euclidean geometry. We will discover that a large part of this geometry is independent of the theory of parallels and is equally valid in hyperbolic geometry.

Review Exercise

Which of the following statements are correct?

(1) The Euclidean parallel postulate states that for every line *l* and for every point P not lying on *l* there exists a unique line *m* through P that is parallel to *l*.

(2) An "angle" is defined as the space between two rays that emanate from a common point.

(3) Most of the results in Euclid's *Elements* were discovered by Euclid himself.

(4) By definitior ⅃ ⅼⅰne *m* is "parallel" to a line *l* if for any two points P, Q on *m,* the perpendicular distance from P to *l* is the same as the perpendicular distance from Q to *l*.

(5) It was unnecessary for Euclid to assume the parallel postulate because the French mathematician Legendre proved it.

(6) A "transversal" to two lines is another line that intersects both of them in distinct points.

(7) By definition, a "right angle" is a 90° angle.

(8) "Axioms" or "postulates" are statements that are assumed, without further justification, whereas "theorems" or "propositions" are proved using the axioms.

(9) We call $\sqrt{2}$ an "irrational number" because it cannot be expressed as a quotient of two whole numbers.

(10) The ancient Greeks were the first to insist on proofs for mathematical statements to make sure they were correct.

Exercises

In Exercises 1–4 you are asked to define some familiar geometric terms. The exercises provide a review of these terms as well as practice in formulating definitions with precision. In making a definition, you may use the five undefined geometric terms and all other geometric terms that have been defined in the text so far or in any preceding exercises.

Making a definition sometimes requires a bit of thought. For example, how would you define *perpendicularity* for two lines *l* and *m*? A first attempt might be to say that "*l* and *m* intersect and at their point of intersection these lines form right angles." It would be legitimate to use the terms "intersect" and "right angle" because they have been previously defined. But what is meant by the statement that *lines* form right angles? Surely, we can all draw a picture to show what we mean, but the problem is to express the idea verbally, using only terms introduced previously. According to the definition on p. 15, an angle is formed by two non-opposite *rays* emanating from the same vertex. We may therefore define *l* and *m* as *perpendicular* if they intersect at a point A and if there is a ray $\overrightarrow{AB}$ that is part of *l* and a ray $\overrightarrow{AC}$ that is part of *m* such that $\not\prec$ BAC is a right angle. We denote this by $l \perp m$.

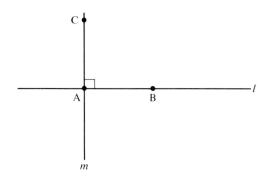

Figure 1.13
Perpendicular lines.

1. Define the following terms:

(a) *Midpoint* M of a segment AB.

(b) *Perpendicular bisector* of a segment AB (you may use the term "midpoint" since you have just defined it).

(c) Ray $\overrightarrow{BD}$ *bisects* angle $\not\prec$ ABC (given that point D is between A and C).

(d) Points A, B, and C are *collinear*.

(e) Lines *l*, *m*, and *n* are *concurrent* (see Figure 1.14).

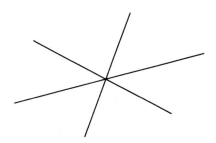

Figure 1.14
Concurrent lines.

2. Define the following terms:

(a) The *triangle* △ABC formed by three noncollinear points A, B, and C.

(b) The *vertices, sides,* and *angles* of △ABC. (The "sides" are segments, not lines.)

(c) The sides *opposite to* and *adjacent to* a given vertex of A of △ABC.

(d) *Medians* of a triangle (see Figure 1.15).

(e) *Altitudes* of a triangle (see Figure 1.16).

(f) *Isosceles* triangle, its *base,* and its *base angles.*

(g) *Equilateral* triangle.

(h) *Right* triangle.

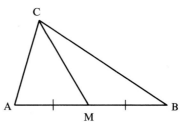

Figure 1.15

Median.

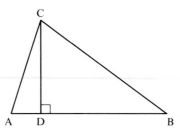

Figure 1.16

Altitude.

3. Given four points A, B, C, and D, no three of which are collinear and such that any pair of the segments AB, BC, CD, and DA either have no point in common or have only an endpoint in common. We can then define the *quadrilateral* □ABCD to consist of the four segments mentioned, which are called its *sides,* the four points being called its *vertices.* (Note that the order in which the letters are written is essential. For example, □ABCD may not denote a quadrilateral, because e.g., AB might cross CD. If □ABCD did denote a quadrilateral, it would not denote the same one as □ACDB. Which permutations of the four letters A, B, C and D do denote the same quadrilateral as □ABCD?) Using this definition, define the following notions:

(a) The *angles* of □ABCD.

(b) *Adjacent* sides of □ABCD.

(c) *Opposite* sides of □ABCD.

(d) The *diagonals* of □ABCD.

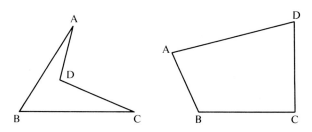

Figure 1.17
Quadrilaterals.

4. Define *vertical angles*. How would you attempt to prove that vertical angles are congruent to each other? (Just sketch a plan for a proof—don't carry it out in detail.)

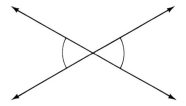

Figure 1.18
Vertical angles.

5. Use a congruence axiom (page 12) to prove the following result: If P and Q are any points on a circle with center O and radius OA, then $OP \cong OQ$.

6. (a) Given two points A and B and a third point C between them. (Recall that "between" is an *undefined* term.) Can you think of any way to prove from the postulates that C lies on line $\overleftrightarrow{AB}$?
   (b) Assuming that you succeeded in proving C lies on $\overleftrightarrow{AB}$, can you prove from the definition of "ray" and the postulates that $\overrightarrow{AB} = \overrightarrow{AC}$?

7. If S and T are any sets, their *union* $(S \cup T)$ and *intersection* $(S \cap T)$ are defined as follows:
   (i) Something belongs to $S \cup T$ if and only if it belongs either to S or to T (or to both of them).
   (ii) Something belongs to $S \cap T$ if and only if it belongs both to S and to T.

   Given two points A and B, consider the two rays $\overrightarrow{AB}$ and $\overrightarrow{BA}$. Draw diagrams to show that $\overrightarrow{AB} \cup \overrightarrow{BA} = \overleftrightarrow{AB}$ and $\overrightarrow{AB} \cap \overrightarrow{BA} = AB$. What additional axioms about the undefined term "between" must we assume in order to be able to *prove* these equalities?

8. To further illustrate the need for careful definition, consider the following possible definitions of "rectangle":

(i) A quadrilateral with four right angles.

(ii) A quadrilateral with all angles congruent to one another.

(iii) A parallelogram with at least one right angle.

In this book we will take (i) *as our definition.* Your experience with Euclidean geometry may lead you to believe that these three definitions are equivalent; sketch informally how you might prove that, and notice carefully which theorems you are tacitly assuming. In hyperbolic geometry these definitions give rise to three different sets of quadrilaterals (see Chapter 6). Given the definition of "rectangle," use it to define "square."

9. Can you think of any way to prove from the postulates that for every line *l*

    (a) there exists a point lying on *l*?
    (b) there exists a point not lying on *l*?

10. Can you think of any way to prove from the postulates that the plane is nonempty, i.e., that points and lines exist? (Discuss with your instructor what it means to say that mathematical objects, such as points and lines, "exist.")

11. Do you think that the Euclidean parallel postulate is "obvious"? Write a brief essay explaining your answer.

12. Do you think the axiomatic method can be applied to subjects other than mathematics? Is the U.S. Constitution (including all its amendments) the list of axioms from which the federal courts logically deduce all rules of law? Do you think the "truths" asserted in the Declaration of Independence are "self-evident"?

13. Write a commentary on the application of the axiomatic method finished in 1675 by Benedict de Spinoza, entitled: *Ethics Demonstrated in Geometrical Order and Divided into Five Parts Which Treat* (1) *of God;* (2) *of the Nature and Origin of the Mind;* (3) *of the Nature and Origin of the Emotions;* (4) *of Human Bondage, or of the Strength of the Emotions;* (5) *of the Power of the Intellect, or of Human Liberty.* (Devote the main body of your review to Parts 4 and 5.)

14. Comment on the following question put to Faust by Mephistopheles: "Is it right, I ask, is it even prudent, to bore thyself and bore thy students?"

Major Exercises

1. In this exercise we will review several basic Euclidean constructions with a straightedge and compass. Such constructions fascinated mathematicians from ancient Greece until the nineteenth century, when all classical construction problems were finally solved.

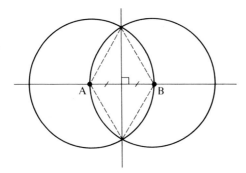

Figure 1.19

(a) Given a segment AB. Construct the perpendicular bisector of AB. (Hint: make AB a diagonal of a rhombus, as in Figure 1.19.)

(b) Given a line l and a point P lying on l. Construct the line through P perpendicular to l. (Hint: make P the midpoint of a segment of l.)

(c) Given a line l and a point P *not* lying on l. Construct the line through P perpendicular to l. (Hint: construct isosceles triangle $\triangle ABP$ with base AB on l and use (a).)

(d) Given a line l and a point P not lying on l. Construct a line through P parallel to l. (Hint: use (b) and (c).)

(e) Construct the bisecting ray of an angle. (Hint: Use the Euclidean theorem that the perpendicular bisector of the base of an isosceles triangle is also the angle bisector of the angle opposite the base.)

(f) Given $\triangle ABC$ and segment $DE \cong AB$. Construct a point F on a given side of line $\overleftrightarrow{DE}$ such that $\triangle DEF \cong \triangle ABC$.

(g) Given angle $\sphericalangle ABC$ and ray $\overrightarrow{DE}$. Construct F on a given side of line $\overleftrightarrow{DE}$ such that $\sphericalangle ABC \cong \sphericalangle FDE$.

2. Euclid assumed the compass to be *collapsible*. That is, given two points P and Q, the compass can draw a circle with center P passing through Q (Postulate III); however, the spike cannot be moved to another center O to draw a circle of the same radius. Once the spike is moved, the compass collapses. Check through your constructions in Exercise 1 to see if they are possible with a collapsible compass. (For purposes of this exercise, being "given" a line means being given two or more points on it.)

(a) Given three points P, Q, and R. Construct with a straightedge and collapsible compass a rectangle $\square PQST$ with PQ as a side and such that $PT \cong PR$ (see Figure 1.20).

(b) Given a segment PQ and a ray $\overrightarrow{AB}$. Construct the point C on $\overrightarrow{AB}$ such that $PQ \cong AC$. (Hint: using (a), construct rectangle $\square PAST$ with $PT \cong PQ$, then draw the circle centered at A and passing through S.)

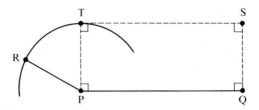

Figure 1.20

Exercise (b) shows that you can transfer segments with a collapsible compass and a straightedge, so you can carry out all constructions as if your compass did not collapse.

3. The straightedge you used in the previous exercises was supposed to be *unruled* (if it did have marks on it, you weren't supposed to use them). Now, however, let us mark two points on the straightedge so as to mark off a certain distance *d*. Archimedes showed how we can then trisect an arbitrary angle:

   For any angle, draw a circle γ of radius *d* centered at the vertex O of the angle. This circle cuts the sides of the angle at points A and B. Place the marked straightedge so that one mark gives a point C on line $\overset{\leftrightarrow}{OA}$ such that O is between C and A, the other mark gives a point D on circle γ, and the straightedge must simultaneously rest on the point B, so that B, C, and D are collinear. Prove that $\measuredangle$ COD so constructed is one-third of $\measuredangle$ AOB. (Hint: use Euclidean theorems on exterior angles and isosceles triangles.)

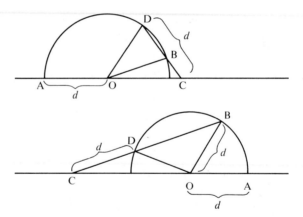

Figure 1.21

4. The number $\rho = (1 + \sqrt{5})/2$ was called the *golden ratio* by the Greeks, and a rectangle whose sides are in this ratio is called a *golden rectangle*.* Prove that a golden rectangle can be constructed with straightedge and compass as follows:

---

* For applications of the golden ratio to Fibonacci numbers and phyllotaxis, see H. M. S. Coxeter's *Introduction to Geometry*, Chapter 11.

(a) Construct a square □ABCD.

(b) Construct midpoint M of AB.

(c) Construct point E such that B is between A and E and MC ≅ ME.

(d) Construct the foot F of the perpendicular from E to $\overset{\leftrightarrow}{DC}$.

(e) Then □AEFD is a golden rectangle (use the Pythagorean theorem for △MBC).

(f) Moreover, □BEFC is another golden rectangle (first show that $1/\rho = \rho - 1$).

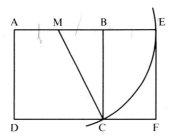

Figure 1.22

The next two exercises require a knowledge of trigonometry.

5. The Egyptians thought that if a quadrilateral had sides of lengths a, b, c, and d, then its area S was given by the formula $(a + c)(b + d)/4$. Prove that actually

$$4S \leqq (a + c)(b + d)$$

with equality holding only for rectangles. (Hint: twice the area of a triangle is $ab \sin \theta$, where θ is the angle between the sides of lengths a, b, and $\sin \theta \leqq 1$, with equality holding only if θ is a right angle.)

6. Prove analogously that if a triangle has sides of lengths a, b, c, then its area S satisfies the inequality

$$4S\sqrt{3} \leqq a^2 + b^2 + c^2$$

with equality holding only for equilateral triangles. (Hint: if θ is the angle between sides b and c, chosen so that it is at most 60°, then use the formulas

$$2S = bc \sin \theta$$
$$2bc \cos \theta = b^2 + c^2 - a^2 \text{ (law of cosines)}$$
$$\cos(60° - \theta) = (\cos \theta + \sqrt{3} \sin \theta)/2$$

The next four exercises require research in the library.

7. Write a paper explaining in detail why it is impossible to trisect an arbitrary angle or square a circle using only a compass and unmarked straightedge (see Eves, Kutuzov, Moise).

8. Here are two other famous results in the theory of constructions:

   (a) The Danish mathematician G. Mohr and the Italian L. Mascheroni discovered independently that all Euclidean constructions of points can be made with a compass alone. A line, of course, cannot be drawn with a compass, but it can be determined with a compass by constructing two points lying on it. In this sense, Mohr and Mascheroni showed that the straightedge is unnecessary.

   (b) On the other hand, the German J. Steiner and the Frenchman J. V. Poncelet showed that all Euclidean constructions can be carried out with a straightedge alone if we are first given a single circle and its center.

   Report on these remarkable discoveries (see Eves, Kutuzov).

9. Given any triangle △ABC. Draw the two rays that trisect each of its angles, and let P, Q, and R be the three points of intersection of adjacent trisectors. Prove Morley's theorem that △PQR is an equilateral triangle (see Figure 1.23 and Coxeter's *Introduction to Geometry*).

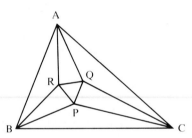

Figure 1.23
Morley's theorem.

10. Here is a research problem whose answer may be known, although I haven't seen it. Is there a nice generalization of Morley's theorem for the cases in which each angle of the triangle is divided into 4, 5, 6, ... equal parts? What if figures other than triangles are used (quadrilaterals, pentagons, hexagons, etc.)?

Reductio ad absurdam . . . is a far finer gambit than any chess gambit: a chess player may offer the sacrifice of a pawn or even a piece, but a mathematician offers the *game.*

G. H. Hardy

Informal Logic

In the previous chapter we were introduced to the postulates and basic definitions of Euclid's geometry, slightly rephrased for greater precision. We would like to begin proving some theorems or propositions that are logical consequences of the postulates. However, the exercises of the previous chapter may have alerted you to expect some difficulties that we must first clear up. For example, there is nothing in the postulates that guarantees that a line has any points lying on it (or off it)! You may feel this is ridiculous—it wouldn't be a line if it didn't have any points lying on it. (What kind of a line is he feeding us anyway?) In a sense, your protest would be legitimate, for if my concept of a line were so different from yours, we would not understand each other, and Requirement 0—that there be mutual understanding of words and symbols used—would be violated.

So let me be perfectly clear. We must play this game according to the rules, the rules mentioned in Requirement 2 but not spelled out. Unfortunately, to discuss them completely would require changing the content of this book from geometry to symbolic logic. Instead, I will only remind you of some basic rules of reasoning that you, as a rational being, already know.

LOGIC RULE 1.   No unstated assumptions may be used in a proof.

The reason for taking the trouble in Chapter 1 to list all our axioms was to be explicit about our basic assumptions, including even the most obvious. Although it is "obvious" that two points determine a unique line, Euclid stated this as his first postulate. So if in some proof we want to say that every line has points lying on it, we should list this state-

Figure 2.1

The shortest path between two points on a sphere is an arc of a *great circle* (a circle whose center is the center of the sphere and whose radius is the radius of the sphere, e.g., the equator).

ment as another postulate (or prove it, but we can't). In other words, all our cards must be out on the table. If you reread Exercises 6, 7, 9 and 10 in Chapter 1, you will find some "obvious" assumptions that we will have to make explicit. This will be done later.

Perhaps you have realized by now that there is a vital relation between axioms and undefined terms. As we have seen, we must have undefined terms in order to avoid infinite regress. But this does not mean we can use these terms in any way we choose. The axioms tell us exactly what properties of undefined terms we are allowed to use in our arguments. You may have some other properties in your mind when you think about these terms, but you're not allowed to use them in a proof (Rule 1). For example, when you think of the unique line determined by two points, you probably think of it as being "straight," or as "the shortest path between the two points." Euclid's postulates do not allow us to assume these properties. Besides, from one viewpoint, these properties could be considered contradictory. If you were traveling the surface of the earth, say from San Francisco to Moscow, the shortest path would be an arc of a *great circle* (a straight path would bore through the earth). Indeed, pilots in a hurry fly their aircraft over great circles.

Theorems and Proofs

All mathematical theorems are conditional statements, statements of the form

If [hypothesis] *then* [conclusion].

In some cases a theorem may state only a conclusion; the axioms of the particular mathematical system are then implicit (assumed) as a hypothesis. If a theorem is not written in the conditional form, it can nevertheless be translated into that form. For example,

> *Base angles of an isosceles triangle are congruent.*

can be interpreted as

> *If a triangle has two congruent sides, then the angles opposite those sides are congruent.*

 ˙ Put another way, a conditional statement says that one condition (the hypothesis) *implies* another (the conclusion). If we denote the hypothesis by H, the conclusion by C, and the word "implies" by an arrow $\Rightarrow$, then every theorem has the form $H \Rightarrow C$. (In the example above, H is "two sides of a triangle are congruent" and C is "the angles opposite those sides are congruent.")

Not every conditional statement is a theorem. For example, the statement

> *If $\triangle ABC$ is any triangle, then it is isosceles.*

is not a theorem. Why not? You might say that this statement is "false" whereas theorems are "true." Let's avoid the loaded words "true" and "false," for they beg the question and lead us into more complicated issues.

In a given mathematical system the only statements we call *theorems** are those statements for which a *proof* has been supplied. We can disprove the assertion that every triangle is isosceles by exhibiting a triangle that is not isosceles, such as a 3-4-5 right triangle.

The crux of the matter then is the notion of *proof*. By definition, a proof is a list of statements, together with a justification for each statement, ending up with the conclusion desired. Usually, each statement in a proof will be numbered in this book, and the justification for it will follow in parentheses. *Only six types of justifications are allowed:*

1. "By hypothesis...."
2. "By axiom...."

---

* Or sometimes *propositions, corollaries,* or *lemmas.* "Theorem" and "proposition" are interchangeable; a "corollary" is an immediate consequence of a theorem, and a "lemma" is a less important result.

3. "By theorem..." (previously proved).
4. "By definition...."
5. "By step..." (a previous step in the argument).
6. "By rule...of logic."

Giving these justifications is the way we observe Rule 1. Later in the the book our proofs will be less formal, and justifications may be omitted when they are obvious. (Be forewarned, however, that these omissions can lead to incorrect results.) A justification may involve several of the above types.

Having described proofs, it would be nice to be able to tell you how to find or construct them. Yet that is the mystery of doing mathematics. Certain techniques for proving theorems are learned by experience, by imitating what others have done. But there is no rote method for proving or disproving every statement in mathematics. (The nonexistence of such a rote method is, when stated precisely, a deep theorem in mathematical logic and is the reason why computers will never put mathematicians out of business—see Delong, Chapter 5).

However, some suggestions may help you construct proofs. First, make sure you clearly understand the meaning of each term in the statement of the proposed theorem. If necessary, review their definitions. Second, keep reminding yourself of what it is you are trying to prove. If it involves parallel lines, for example, look up previous propositions that give you information about parallel lines. If you find another proposition that seems to apply to the problem at hand, check carefully to see whether it really does apply. Draw pictures to help you visualize the problem.

RAA Proofs

The most common type of proof in this book is proof by reductio ad absurdum, abbreviated RAA. In this type of proof you want to prove a conditional statement, $H \Rightarrow C$, and you begin by assuming the opposite of the conclusion you seek. We call this opposite assumption the *RAA hypothesis*, to distinguish it from the hypothesis H. The RAA hypothesis is a temporary assumption from which we derive, by reasoning, an *absurd statement* ("absurd" in the sense that it denies something known to be valid). Such a statement might deny the hypothesis of the theorem or the RAA hypothesis; it might deny a previously proved theorem or an axiom. Once it is shown that the negation of C leads to an absurdity, it follows that C must be valid. This is called the *RAA conclusion*. To summarize:

LOGIC RULE 2.   To prove a statement $H \Rightarrow C$, assume the negation of statement C (RAA hypothesis) and deduce an absurd statement, using the hypothesis H in your deduction.

Let us illustrate this rule by proving the following proposition (Proposition 2.1): If l and m are distinct lines that are not parallel, then l and m have a unique point in common.

PROOF:
(1) Because l and m are not parallel, they have a point in common (by definition of "parallel").
(2) Since we want to prove uniqueness for the point in common, we will assume the contrary, that l and m have two distinct points A and B in common (RAA hypothesis).
(3) Then there is more than one line on which A and B both lie [step (2) and the hypothesis of the theorem, $l \neq m$].
(4) Step (3) contradicts Euclid's first postulate.
(5) Hence, l and m have a unique point in common (RAA conclusion).■

As another illustration, consider one of the earliest RAA proofs, discovered by the Pythagoreans (to their great dismay). In giving this proof, we will use some facts about Euclidean geometry and numbers that you know, and we will be informal.

Suppose $\triangle$ABC is a right isosceles triangle with right angle at C. We can choose our unit of length so that the legs have length one. The theorem then says that the length of the hypotenuse is irrational.

By the Pythagorean theorem, the length of the hypotenuse is $\sqrt{2}$, so we must prove that $\sqrt{2}$ is an irrational number, i.e., that it is not a rational number.

What is a rational number? It is a number that can be expressed as a quotient p/q of two whole numbers p and q. For example, $\frac{1}{2}$, $\frac{2}{3}$, and $5 = \frac{5}{1}$ are rational numbers. We want to prove that $\sqrt{2}$ is not one of these numbers.

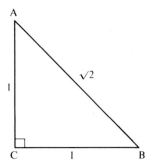

Figure 2.2

We begin by assuming the contrary, that $\sqrt{2}$ *is* a rational number (RAA hypothesis). In other words, $\sqrt{2} = p/q$ for certain unspecified whole numbers p and q. You know that every rational number can be written in lowest terms, i.e., such that the numerator and denominator have no common factor. For example, $\frac{4}{6}$ can be written as $\frac{2}{3}$, where the common factor 2 in the numerator and denominator has been cancelled. Thus we can assume all common factors have been cancelled, so that p and q have no common factor.

Next, we clear denominators

$$\sqrt{2}q = p$$

and square both sides

$$2q^2 = p^2.$$

This equation says that p^2 is an even number (since p^2 is twice another whole number, namely q^2). If p^2 is even, p must be even, for the square of an odd number is odd, as you know. Thus,

$$p = 2r$$

for some whole number r (that is what it means to be even). Substituting $2r$ for p in the previous equations gives

$$2q^2 = (2r)^2 = 4r^2.$$

We then cancel 2 from both sides to get

$$q^2 = 2r^2.$$

This equation says that q^2 is an even number, hence q must be even.

We have shown that numerator p and denominator q are both even, meaning that they have 2 as a common factor. Now this is absurd, because all common factors were cancelled. Thus, $\sqrt{2}$ is irrational (RAA conclusion).

Negation

In an RAA proof we begin by "assuming the contrary." Sometimes the contrary or negation of a statement is not obvious, so you should know the rules for negation.

First, some remarks on notation. If S is any statement, we will denote the negation or contrary of S by $\sim S$. For example, if S is the statement "p is even," then $\sim S$ is the statement "p is not even" or "p is odd."

The rule below applies to those cases where S is already a negative statement. The rule states that two negatives make a positive.

LOGIC RULE 3. The statement "$\sim(\sim S)$" means the same as "S."

We followed this rule when we negated the statement "$\sqrt{2}$ is irrational" by writing the contrary as "$\sqrt{2}$ is rational" instead of "$\sqrt{2}$ is not irrational."

Another rule we have already followed in our RAA method is the rule for negating an implication. We wish to prove $H \Rightarrow C$, and we assume, on the contrary, H does not imply C, i.e., that H holds and at the same time $\sim C$ holds. We write this symbolically as H & $\sim C$, where & is the abbreviation for "and." A statement involving the connective "and" is called a *conjunction*. Thus,

LOGIC RULE 4. The statement "$\sim[H \Rightarrow C]$" means the same as "H & $\sim C$."

Let us consider, for example, the conditional statement "if 3 is an odd number, then 3^2 is even." According to Rule 4, the negation of this is the declarative statement "3 is an odd number and 3^2 is odd."

How do we negate a conjunction? A conjunction S_1 & S_2 means that statements S_1 and S_2 both hold. Negating this would mean asserting that one of them does not hold, i.e., asserting the negation of one or the other. Thus

LOGIC RULE 5.   The statement "$\sim[S_1 \ \& \ S_2]$" means the same as "$[\sim S_1 \text{ or } \sim S_2]$."

A statement involving the connective "or" is called a *disjunction*. The mathematical "or" is not exclusive like "or" in everyday usage. Consider the conjunction "$1 = 2$ and $1 = 3$." If we wish to deny this, we must write (according to Rule 5) "$1 \neq 2$ or $1 \neq 3$." Of course, both inequalities are valid. So when a mathematician writes "S_1 or S_2" he means "either S_1 holds or S_2 holds *or they both hold*."

Finally let us be more precise about what is an absurd statement. It is the conjunction of a statement S with the negative of S, i.e., "$S \ \& \sim S$." A statement of this type is called a *contradiction*. A system of axioms from which no contradiction can be deduced is called *consistent*.

Quantifiers

Most mathematical statements involve *variables*. For instance, the Pythagorean theorem states that for any right triangle, if a and b are the lengths of the legs and c the length of the hypotenuse, then $c^2 = a^2 + b^2$. Here a, b, and c are variable numbers, and the triangle whose sides they measure is a variable triangle.

Variables can be quantified in two different ways. First, in a *universal* way, as in the expressions:
"For any x, ..."
"For every x, ..."
"For all x, ..."
"Given any x, ..."
"If x is any . . . ."

Second, in an *existential* way, as in the expressions:

"For some x, ..."
"There exists an x . . . ."
"There is an x . . . ."
"There are x . . . ."

Consider Euclid's first postulate, which states informally that two points P and Q determine a unique line *l*. Here P and Q may be any two points, so they are quantified universally, whereas l is quantified existentially, since it is asserted to exist, once P and Q are given.

It must be emphasized that a statement beginning with "For every..." does not imply the existence of anything. The statement "every unicorn has a horn on its head" does not imply that unicorns exist.

If a variable x is quantified universally, this is usually denoted as $\forall x$, (read as "for all x"). If x is quantified existentially, this is usually denoted as $\exists x$ (read as "there exists an x..."). After a variable x is quantified, some statement is made about x, which we can write as $S(x)$ (read as "statement S about x"). Thus, a universally quantified statement about a variable x has the form $\forall x S(x)$.

We wish to have rules for negating quantified statements. How do we deny that statement $S(x)$ holds for all x? We can do so clearly by asserting that for some x, $S(x)$ does not hold.

LOGIC RULE 6.   The statement "$\sim [\forall x S(x)]$" means the same as "$\exists x \sim S(x).$"

For example, to deny "All triangles are isosceles" is to assert "There is a triangle that is not isosceles."

Similarly, to deny that there exists an x having property $S(x)$ is to assert that all x fail to have property $S(x)$.

LOGIC RULE 7.   The statement "$\sim [\exists x S(x)]$" means the same as "$\forall x \sim S(x).$"

For example, to deny "There is an equilateral right triangle" is to assert "Every right triangle is nonequilateral" or, equivalently, to assert "No right triangle is equilateral."

Since in practice quantified statements involve several variables, the above rules will have to be applied several times. Usually, common sense will quickly give you the negation. If not, follow the above rules.

Let's work out the denial of Euclid's first postulate. This postulate is a statement about all pairs of points P and Q; negating it would mean, according to Rule 6, asserting the existence of points P and Q that do not satisfy the postulate. Postulate I involves a conjunction, asserting that P and Q lie on some line l *and* that l is unique. In order to deny this conjunction, we follow Rule 5. The assertion becomes either "P and Q do not lie on any line" *or* "they lie on more than one line." Thus, the negation of Postulate I asserts: "There are two points P and Q that either do not lie on any line or that lie on more than one line."

If we return to the example of the surface of the earth, thinking of a "line" as a great circle, we see that there do exist such points P and Q— namely, take P to be the north pole and Q the south pole. Infinitely many great circles pass through both poles.

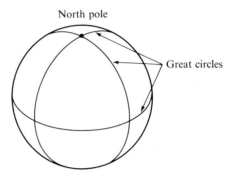

North pole

Great circles

Figure 2.3

Mathematical statements are sometimes made informally, and you may sometimes have to rephrase them before you will be able to negate them. For example, consider the following statement:

If a line intersects one of two parallel lines, it also intersects the other.

This appears to be a conditional statement, of the form "if...then..."; its negation, according to Rule 4, would appear to be:

A line intersects one of two parallel lines and does not intersect the other.

If this seems awkward, it is because the original statement contained *hidden quantifiers* that have been ignored. The original statement refers to *any* line that intersects one of two parallel lines, and these are *any* parallel lines. There are universal quantifiers implicit in the original statement. So we have to follow Rule 6 as well as Rule 4 in forming the correct negation, which is:

There exist two parallel lines and a line that intersects one of them and does not intersect the other.

Implication

Another rule, called the *rule of detachment,* or *modus ponens,* is the following:

LOGIC RULE 8.   If $P \Rightarrow Q$ and P are steps in a proof, then Q is a justifiable step.

This rule is almost a definition of what we mean by implication. For example, we have an axiom stating that if $\angle$ A and $\angle$ B are right angles, then $\angle$ A $\cong$ $\angle$ B (Postulate IV). Now in the course of a proof we may come across two right angles. Rule 8 allows us to assert their congruence as a step in the proof.

You should beware of confusing a conditional statement $P \Rightarrow Q$ with its *converse* $Q \Rightarrow P$. For example, the converse of Postulate IV states that if $\angle$ A $\cong$ $\angle$ B then $\angle$ A and $\angle$ B are right angles, which is absurd.

However, it may sometimes happen that both a conditional statement and its converse are valid. In case $P \Rightarrow Q$ and $Q \Rightarrow P$ both hold, we write simply $P \Leftrightarrow Q$ (read as *"P if and only if Q"* or *"P is logically equivalent to Q"*). All definitions are of this form. For example, three points are collinear if and only if they lie on a line. Some theorems are also of this form, such as the theorem "a triangle is isosceles if and only if two of its angles are congruent to each other." The next rule gives a few more ways that "implication" is often used in proofs.

LOGIC RULE 9.   (a) $P \Rightarrow [P \text{ or } Q]$, $Q \Rightarrow [P \text{ or } Q]$.
(b) $[P \,\&\, Q] \Rightarrow P$, $[P \,\&\, Q] \Rightarrow Q$.
(c) $[\sim Q \Rightarrow \sim P] \Leftrightarrow [P \Rightarrow Q]$.
(d) $\big[[P \Rightarrow Q] \,\&\, [Q \Rightarrow R]\big] \Rightarrow [P \Rightarrow R]$.

Law of Excluded Middle

LOGIC RULE 10.   For every statement P, either P holds or its negation $\sim P$ holds.

In other words, assertions of the form "P or $\sim P$" are valid steps in a proof. For example, given two lines, we may assert that either they intersect or they are parallel.

The law of the excluded middle has provoked much controversy in mathematics and philosophy. Mathematicians of the intuitionist school (Brouwer, Heyting, Bishop) object to the unqualified use of this rule when statements about existence are involved. They contend that in order to prove that a mathematical object exists, you must first supply an effective method for constructing it. According to intuitionists, it is not sufficient to assume that the object does not exist (RAA hypothesis) and then derive a contradiction.

The law of the excluded middle is characteristic of *two-valued logic*: either a statement holds or it does not; there is no middle ground. This has sometimes been described as "God's logic"—even if we mortals can't tell whether a statement is valid, God knows. In recent years, research has begun on *many-valued logic* (Lukasiewicz, Post, Tarski), but this work has so far not had much impact on the mainstream of mathematics.

Incidence Geometry

Let us apply the logic we have developed to a very elementary part of geometry, *incidence geometry*. We assume only the undefined terms "point" and "line" and the undefined relation "incidence" between a point and a line, written "P lies on *l*" as before. We don't discuss "betweenness" or "congruence" in this geometry.

These undefined terms will be subjected to three axioms, the first of which is the same as Euclid's first postulate.

INCIDENCE AXIOM 1.   For every point P and for every point Q not equal to P there exists a unique line *l* that passes through* P and Q.

INCIDENCE AXIOM 2.   For every line *l* there exist at least two distinct points incident with *l*.

INCIDENCE AXIOM 3.   There exist three distinct points with the property that no line is incident with all three of them.

* "Passes through" is another way of saying "is incident with."

These axioms fill the gap mentioned in Exercises 9 and 10, Chapter 1. We can now assert that every line has points lying on it—at least two, possibly more—and that the points do not all lie on one line. Moreover, we know that the geometry must have at least three points in it, by the third axiom and Rule 9b of logic. Namely, Incidence Axiom 3 is a conjunction of two statements:

1. There exist distinct points A, B, and C.
2. For every line, at least one of these points does not lie on the line.

Rule 9b tells us that a conjunction of two statements implies each statement separately, so we can conclude that three distinct points exist.

Incidence geometry has some defined terms, such as "collinear," "concurrent," and "parallel," defined exactly as they were in Chapter 1. Incidence Axiom 3 can be rewritten as "there exist three noncollinear points." Parallel lines are still lines that do not have a point in common.

What sort of results can we prove using this meager collection of axioms? None that are very exciting, but here are a few you can prove as exercises.

PROPOSITION 2.1.   If *l* and *m* are distinct lines that are not parallel, then *l* and *m* have a unique point in common.

PROPOSITION 2.2.   For every line there is at least one point not lying on it.

PROPOSITION 2.3   For every point there is at least one line not passing through it.

PROPOSITION 2.4.   For every point P there exist at least two distinct lines that pass through P.

PROPOSITION 2.5.   There exist three distinct lines that are not concurrent (i.e., such that no one point lies on all three lines).

Models

In reading over the axioms of incidence in the previous section, you may have imagined dots and long dashes drawn on a sheet of paper. With this representation in mind, the axioms appear to be correct statements. We will take the point of view that these dots and dashes are a *model* for incidence geometry.

More generally, if we have any axiom system, we can interpret the undefined terms in some way, i.e., give the undefined terms a particular meaning. We call this an *interpretation* of the system. We can then ask whether the axioms, so interpreted, are correct statements. If they are, we call the interpretation a *model*. When we take this point of view, interpretations of the undefined terms "point," "line," and "incident" other than the usual dot-and-dash drawings become possible.

Example 1. Consider a set {A, B, C} of three letters, which we will call "points." "Lines" will be those subsets that contain exactly two letters— {A, B}, {A, C}, and {B, C}. A "point" will be interpreted as "incident" with a "line" if it is a member of that subset. Thus, under this interpretation, A lies on {A, B} and {A, C} but does not lie on {B, C}. In order to determine whether this interpretation is a *model,* we must check whether the interpretations of the axioms are correct statements. For Incidence Axiom 1, if P and Q are any two of the letters A, B, or C, {P, Q} is the unique "line" on which they both lie. For Axiom 2, if {P, Q} is any "line," P and Q are two distinct "points" lying on it. For Axiom 3, we see that A, B, and C are three distinct "points" that are not collinear.

What is the use of models? The main property of any model of an axiom system is that all theorems of the system are correct statements in the model. This is because logical consequences of correct statements are themselves correct. (By definition of "model," axioms are correct statements when interpreted in models; theorems are logical consequences of axioms.) Thus, we immediately know that the five propositions in the previous section hold in the three-point geometry above (Example 1).

Suppose we have a statement in the formal system but don't yet know whether it is a theorem, i.e., we don't yet know whether it can be proved. We can look at our models and see whether the statement is correct in the models. If we can find *one* model where the interpreted statement fails to hold, we can be sure that no proof is possible. You are undoubtedly familiar with testing for the correctness of geometric statements by drawing pictures. Of course, the converse does not work; just because a drawing makes a statement *look* right does not mean you can *prove* it. This will be illustrated in the next section.

The advantage of having several models is that a statement may hold in one model but not in another. Models are "laboratories" for experimenting with the formal system.

Let us experiment with the Euclidean parallel postulate. This is a statement in the formal system incidence geometry: "For every line *l* and every point P not lying on *l* there exists a unique line through P that is parallel to *l*." This statement appears to be correct according to our drawings (although we cannot verify the uniqueness of the parallelism, since we cannot extend our dashes indefinitely). But what about our three-point model? It is immediately apparent that *no parallel lines exist* in this model: {A, B} meets {B, C} in the point B and meets {A, C} in the point A; {B, C} meets {A, C} in the point C. (We say that this model has the *elliptic parallel property*.)

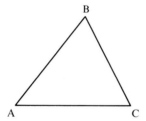

Figure 2.4

Elliptic parallel property (no parallel lines). A 3-point incidence geometry.

Thus, we can conclude that *no proof of the Euclidean parallel postulate from the axioms of incidence alone is possible; in fact, in incidence geometry it is impossible to prove that parallel lines exist.* Similarly, the contrary statement "any two lines have a point in common" (the elliptic parallel property) cannot be proved from the axioms of incidence geometry, for if you could prove it, it would hold in the usual drawn model (and in the models in Examples 3 and 4 below).

The technical description for this situation is that the statement "parallel lines exist" is "independent" of the axioms of incidence. We call a statement *independent* of given axioms if it is impossible to either prove or disprove the statement from the axioms. Independence is demonstrated by constructing two models for the axioms: one in which the statement holds and one in which it does not hold. This method will be used very decisively in Chapter 7 to settle once and for all the question of whether the parallel postulate can be proved.

An axiom system is called *complete* if there are no independent statements in the language of the system, i.e., every statement in the language of the system can either be proved or disproved from the axioms. Thus, the axioms for incidence geometry are incomplete. The axioms for Euclidean and hyperbolic geometries given later in the book can be proved to be complete (see Tarski's article in *The Axiomatic Method,* Henkin, Suppes and Tarski, eds.).

Example 2. Suppose we interpret "points" as points on a sphere, "lines" as great circles on the sphere, and "incidence" in the usual sense, as a point lying on a great circle. In this interpretation there are again no parallel lines. However, this interpretation is *not* a model for incidence geometry, for, as was already mentioned, the interpretation of Incidence Axiom 1 fails to hold—there are an infinite number of great circles passing through the north and south poles on the sphere (see Figure 2.3).

Example 3. Let the "points" be the four letters A, B, C, and D. Let the "lines" be all six sets containing exactly two of these letters: {A, B}, {A, C}, {A, D}, {B, C}, {B, D}, and {C, D}. Let "incidence" be set membership, as in Example 1. As an exercise, you can verify that this is a model for incidence geometry and that in this model the Euclidean parallel postulate does hold (see Figure 2.5).

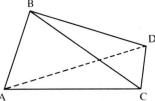

Figure 2.5

Euclidean parallel property. A 4-point incidence geometry.

Example 4. Let the "points" be the five letters A, B, C, D, and E.* Let the "lines" be all ten sets containing exactly two of these letters.

---

* An incidence geometry with only finitely many points is called a *finite geometry.* There is an entertaining discussion of finite geometries (with applications to growing tomato plants) in Chapter 4 of Beck, Bleicher and Crowe. For an advanced treatment, see Dembowski or Stevenson. See the exercises at the end of this chapter for more examples.

Let "incidence" be set membership, as in Examples 1 and 3. You can verify that in this model the following statement about parallel lines, characteristic of hyperbolic geometry, holds: "For every line *l* and every point P not on *l* there exist at least two lines through P parallel to *l*." (See Figure 2.6.)

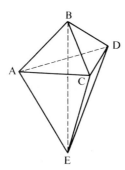

Figure 2.6

Hyperbolic parallel property. A 5-point incidence geometry.

Let us summarize the significance of models. Models can be used to prove the independence of a statement from given axioms, i.e., models can be used to demonstrate the impossibility of proving or disproving a statement from the axioms. Moreover, if an axiom system has many models that are essentially different from each other,* as the models in Examples 1, 3, and 4 are essentially different from each other, then that system has a wide range of applicability. Propositions proved from the axioms of such a system are automatically correct statements within *any* of the models. Mathematicians often discover that an axiom system constructed with one particular model in mind has applications to completely different models never dreamed of.

At the other extreme, when all models of an axiom system are isomorphic to one another, the axioms are called *categorical*. (The axioms for Euclidean and hyperbolic geometries given later in the book are categorical.) The advantage of categorical axioms is that they completely describe all properties of the model that are expressible in the language of the system. In other words, categorical axioms are complete. (For a simple example of a categorical system, suppose we add to the three incidence axioms a fourth axiom asserting that there do not exist four

---

* The technical expression is *nonisomorphic* to each other—see Exercise 10.

distinct points. Obviously, the three-point model in Example 1 is the only model, up to isomorphism, for this expanded axiom system.)

Finally, models provide evidence for the consistency of the axiom system. For instance, incidence geometry must be consistent, since we are unable to deduce any contradiction about the three letters A, B, and C (Example 1).

The Danger in Diagrams

Diagrams have always been helpful in understanding geometry—they are included in Euclid's *Elements* and they are included in this book. But there is a danger that a diagram may suggest a fallacious argument. A diagram may be slightly inaccurate or it may represent only a special case. If we are to recognize the flaws in arguments such as Legendre's (Chapter 1, pp. 19–21), we must not be misled by diagrams that *look* plausible.

What follows is a well-known and rather involved argument that pretends to prove that all triangles are isosceles. For the present, stop thinking from first principles and place yourself in the context of high-school geometry just for purposes of this section. (After this section you will be expected to play dumb again.) See if you can find the flaw in the argument.

Given $\triangle ABC$. Construct the bisector of $\measuredangle A$ and the perpendicular bisector of side BC opposite to $\measuredangle A$. Consider the various cases.

Case 1. The bisector of $\measuredangle A$ and the perpendicular bisector of segment BC are either parallel or identical. In either case, the bisector of $\measuredangle A$ is perpendicular to BC and hence, by definition, is an altitude. Therefore, the triangle is isosceles. (The conclusion follows from the Euclidean theorem: if an angle bisector and altitude from the same vertex of a triangle coincide, the triangle is isosceles.)

Suppose now that the bisector of $\measuredangle A$ and the perpendicular bisector of the side opposite are not parallel and do not coincide. Then they intersect in exactly one point, D, and there are three cases to consider:

Case 2. The point D is inside the triangle
Case 3. The point D is on the triangle.
Case 4. The point D is outside the triangle.

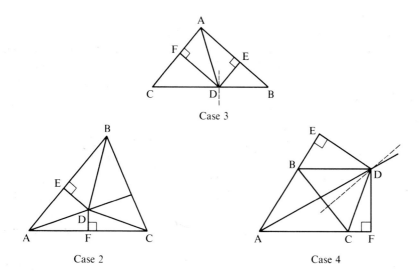

Case 3

Case 2

Case 4

Figure 2.7

For each case construct DE perpendicular to AB and DF perpendicular to AC, and for Cases 2 and 4 join D to B and D to C. In each case, the following proof now holds (see Figure 2.7):

DE ≅ DF because all points on an angle bisector are equidistant from the sides of the angle; DA ≅ DA, and ⦨ DEA and ⦨ DFA are right angles; hence, △ADE is congruent to △ADF by the hypotenuse-leg theorem of Euclidean geometry. (We could also have used the SAA theorem with DA ≅ DA, and the bisected angle and right angles.) Therefore, we have AE ≅ AF. Now, DB ≅ DC because all points on the perpendicular bisector of a segment are equidistant from the ends of the segment. Also, DE ≅ DF, and ⦨ DEB and ⦨ DFC are right angles. Hence, △DEB is congruent to △DFC by the hypotenuse-leg theorem, hence FC ≅ BE. It follows that AB ≅ AC—in Cases 2 and 3 by addition and in Case 4 by subtraction. The triangle is therefore isosceles.

To summarize the proof: it has been shown that if an angle bisector and the perpendicular bisector of the opposite side coincide or are parallel, the triangle is isosceles. It is known that if they do not coincide and are not parallel, they must intersect in exactly one point that is either inside, on, or outside the triangle. It has been shown that in every such case the triangle is isosceles. Therefore, every possible case has been considered and the theorem is proved.

Review Exercise

Which of the following statements are correct?

(1) The "hypothesis" of a theorem is an assumption that implies the conclusion.

(2) A theorem may be proved by drawing an accurate diagram.

(3) To say that a step is "obvious" is an allowable justification in a rigorous proof.

(4) There is no way to program a computer to prove or disprove every statement in mathematics.

(5) To "disprove" a statement means to prove the negation of that statement.

(6) A "model" of an axiom system is the same as an "interpretation" of the system.

(7) The Pythagoreans discovered the existence of irrational lengths by an RAA proof.

(8) The negation of the statement "If 3 is an odd number, then 9 is even" is the statement "If 3 is an odd number, then 9 is odd."

(9) The negation of a conjunction is a disjunction.

(10) The statement "$1 = 2$ and $1 \neq 2$" is an example of a contradiction.

(11) The statement "Base angles of an isosceles triangle are congruent" has no hidden quantifiers.

(12) The statements "Some triangles are equilateral" and "There exists an equilateral triangle" have the same meaning.

(13) The converse of the statement "If you push me, then I will fall" is the statement "If you push me, then I won't fall."

(14) The following two statements are logically equivalent: "If $l \| m$, then l and m have no point in common." "If l and m have a point in common, then l and m are not parallel."

(15) Whenever a conditional statement is valid, its converse is also valid.

(16) If one statement implies a second statement, and the second statement implies a third statement, then the first statement implies the third statement.

(17) The negation of "All triangles are isosceles" is "No triangles are isosceles."

(18) The hyperbolic parallel property is defined as "For every line l and every point P not on l there exist at least two lines through P parallel to l."

(19) The statement "Every point has at least two lines passing through it" is independent of the axioms for incidence geometry.

Exercises

1. What is the flaw in the "proof" (pp. 48–49) that all triangles are isosceles? (All the theorems from Euclidean geometry used in the argument are correct.)

2. (a) What is the negation of $[P \text{ or } Q]$?
   (b) What is the negation of $[P \ \& \ \sim Q]$?
   (c) Using the rules of logic given in the text, show that $P \Rightarrow Q$ means the same as $[\sim P \text{ or } Q]$. (Hint: show they are both negations of the same thing.)
   (d) A symbolic way of writing Rule 2 for RAA proofs is $\big[[H \ \& \ \sim C] \Rightarrow [S \ \& \ \sim S]\big] \Rightarrow [H \Rightarrow C]$. Explain this.

3. Negate Euclid's fourth postulate.

4. Negate the Euclidean parallel postulate.

5. Write out the converse to the following statements:
   (a) "If lines l and m are parallel, then a transversal t to lines l and m cuts out congruent alternate interior angles."
   (b) "If the sum of the degree measures of the interior angles on one side of transversal t is less than $180°$, then lines l and m meet on that side of transversal t."

6. Prove all five propositions in incidence geometry (p. 43). Don't use Incidence Axiom 2 in your proofs.

7. For any pair of axioms of incidence geometry, construct an interpretation in which those two axioms only are satisfied but the third axiom is not. (This will show that the three axioms are *independent,* in the sense that it is impossible to prove any one of them from the other two.)

8. Prove that the interpretations in Examples 3 and 4 (pp. 46–47) are models of incidence geometry in which the Euclidean and hyperbolic parallel properties, respectively, hold.

9. In each of the following interpretations of the undefined terms, which of the axioms of incidence geometry are satisfied and which are not? Tell whether each interpretation has the elliptic, Euclidean, or hyperbolic parallel property.

(a) "Points" are dots on a sheet of paper, "lines" are circles drawn on the paper, "incidence" means that the dot lies on the circle.

(b) Given three letters a, b, and c. Call these letters "lines." Call the two-element sets {a, b}, {a, c}, and {b, c} "points." Let "incidence" be set membership, e.g., "point" {a, b} is incident with "line" a and "line" b but not with "line" c.

(c) "Points" are lines in Euclidean three-dimensional space, "lines" are planes in Euclidean three-space, "incidence" is the usual relation of a line lying in a plane.

(d) Same as in (c), except that we restrict ourselves to lines and planes that pass through a fixed ordinary point O.

(e) Fix a circle in the Euclidean plane. Interpret "point" to mean an ordinary Euclidean point *inside* the circle, interpret "line" to mean a chord of the circle, and let "incidence" mean that the point lies on the chord in the usual sense. (A *chord* of a circle is a segment whose end-points lie on the circle.)

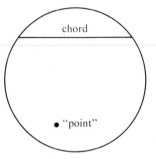

Figure 2.8

(f) Fix a sphere in Euclidean three-space. Two points on the sphere are called *antipodal* if they lie on a diameter of the sphere, e.g., the north and south poles are antipodal. Interpret a "point" to be a set {P, P'} consisting of two antipodal points on the sphere. Interpret a "line" to be a great circle C on the sphere. Interpret a "point" {P, P'} to "lie on" a "line" C if one of the points P, P' lies on the great circle C (then the other point also lies on C).

10. Two interpretations of incidence geometry are called *isomorphic* if a one-to-one correspondence can be set up between the "points" in one interpretation and the "points" in the other, and between the "lines" in one and the "lines" in the other—say, "point" P corresponds to "point" P', "line" *l* corresponds to "line" *l'*—such that the incidence relation is preserved by the correspondence (i.e., P lies on *l* if and only if P' lies on *l'*). For example, the three-point interpretation in Example 1 (p. 44) is isomorphic to the three-point interpretation in Exercise 9b above, using, for example, the following correspondences:

$$A \leftrightarrow \{a, b\} \qquad \{A, B\} \leftrightarrow b$$
$$B \leftrightarrow \{b, c\} \qquad \{B, C\} \leftrightarrow c$$
$$C \leftrightarrow \{a, c\} \qquad \{A, C\} \leftrightarrow a$$

On the other hand, if we use a different correspondence for the "lines," such as

$$\{A, B\} \leftrightarrow a$$
$$\{B, C\} \leftrightarrow b$$
$$\{A, C\} \leftrightarrow c$$

this does not preserve incidence—e.g., A lies on $\{A, C\}$ but the corresponding "point" $\{a, b\}$ to A does not lie on the corresponding "line" c—so this would not be an *isomorphism*. As you can see from this example, the idea is that isomorphic models are essentially the same, only the notation is different.

(a) Prove that when each of two models of incidence geometry has exactly three "points" in it, the models must be isomorphic.

(b) Suppose that two models of incidence geometry have different parallel properties, e.g., one is Euclidean and one is hyperbolic, or one is elliptic and one is Euclidean, etc. Prove that the two models cannot be isomorphic.

(c) Must two models having exactly four "points" be isomorphic? If you think so, prove this assertion; if you think not, give a counter-example.

(d) Show that the models in Exercises 9d and 9f are isomorphic. (Hint: take the point O of Exercise 9d to be the center of the sphere in Exercise 9f, and cut the sphere with lines and planes through O.)

11. Construct a model of incidence geometry that has neither the elliptic, hyperbolic, nor Euclidean parallel properties. (These properties refer to any line *l* and any point P not on *l*. Construct a model that has different parallelism properties for different choices of *l* and P.)

12. Suppose that in a given model for incidence geometry every "line" has at least three distinct "points" lying on it. What are the least number of "points" and the least number of "lines" such a model can have? Suppose further that the model has the Euclidean parallel property. Show that 9 is now the least number of "points" and 12 the least number of "lines" such a model can have.

13. The following syllogisms are by Lewis Carroll. Which of them are correct arguments?

(a) No frogs are poetical; some ducks are unpoetical. Hence, some ducks are not frogs.

(b) Gold is heavy; nothing but gold will silence him. Hence, nothing light will silence him.

(c) All lions are fierce; some lions do not drink coffee. Hence, some creatures that drink coffee are not fierce.

(d) Some pillows are soft; no pokers are soft. Hence, some pokers are not pillows.

14. Consider the following statement, which we will call S: "The statement S is false." Show that if S is either true or false, then there is a contradiction in our

language. (This exercise reveals one philosophical difficulty in developing a precise theory of "truth"—see the *Liar paradox* in the index to DeLong.)

15. Comment on the following statement by the artist David Hunter: "The only use for Logic is writing books on Logic and teaching courses in Logic; it has no application to human behavior."

Major Exercises

1. Let $\triangle ABC$ be such that AB is not congruent to AC. Let D be the point of intersection of the bisector of ⨟ A and the perpendicular bisector of side BC. Let E, F, and G be the feet of the perpendiculars dropped from D to $\overleftrightarrow{AB}$, $\overleftrightarrow{AC}$, $\overleftrightarrow{BC}$, respectively. Prove that

   (a) D lies outside the triangle on the circle through ABC;
   (b) one of E or F lies inside the triangle and the other outside;
   (c) E, F, and G are collinear.

   (This is an exercise in Euclidean geometry. Use anything you know, including coordinates if necessary.)

2. Let M be a model for incidence geometry that has the elliptic parallel property. Define a new interpretation M* by taking as "points" of M* the "lines" of M, as "lines" of M* the "points" of M, with the same incidence relation. Prove that M* is also a model with the elliptic parallel property.

   If, in addition to the elliptic parallel property, M has the property that every line has *at least three* points lying on it, then M is called *a projective plane.* Prove that in a projective plane M with a finite number of points all the lines in M have the same number of points lying on them, and prove that M* is also a finite projective plane.

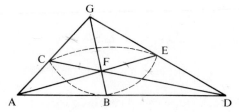

Figure 2.9
The smallest projective plane (7 points).

3. Let us add to the axioms of incidence geometry the following axioms:

   (i) The Euclidean parallel property.
   (ii) The existence of only a finite number of points.

(iii) The existence of lines l and m such that the number of points lying on l is different from the number of points lying on m.

Show that this expanded axiom system is inconsistent. (Hint: prove that (i) and (ii) imply the negation of (iii).)

4. A model for incidence geometry having the Euclidean parallel property is also called an *affine plane*.

   (a) Given an affine plane, add to it one new line called the *line at infinity*. For each family of ordinary lines consisting of all the lines parallel to a particular line in the family add a *point at infinity* that lies on all these parallel lines and on the line at infinity. Prove that this expanded system is a projective plane called the *projective completion* of the given affine plane.

   (b) Prove that every projective plane is isomorphic to the projective completion of an affine plane. (Hint: pick any line and call it "the line at infinity.")

   (c) Apply Exercise 2 and Exercise 4a to obtain another solution to Exercise 3.

5. Consider the following interpretation of incidence geometry: begin with a punctured sphere in Euclidean three-space, i.e., a sphere with one point N removed. Interpret "points" as points on the punctured sphere. For each circle on the original sphere passing through N, interpret the punctured circle obtained by removing N as a "line." Interpret "incidence" in the Euclidean sense of a point lying on a punctured circle. Is this interpretation a model? If so, what parallel property does it have? Is it isomorphic to any other model you know? (Hint: use stereographic projection.)

6. Consider the following statement in incidence geometry: "For any two lines l and m there exists a one-to-one correspondence between the set of points lying on l and the set of points lying on m." Prove that this statement is independent of the axioms of incidence geometry.

7. Let M be a finite projective plane so that, according to Major Exercise 2, all lines in M have the same number of points lying on them; call this number $n + 1$. Prove the following:

   (a) Each point in M has $n + 1$ lines passing through it.
   (b) The total number of points in M is $n^2 + n + 1$.
   (c) The total number of lines in M is $n^2 + n + 1$.

8. Let A be a finite affine plane so that, according to Major Exercise 3, all lines in A have the same number of points lying on them; call this number n. Prove the following:

   (a) Each point in A has $n + 1$ lines passing through it.
   (b) The total number of points in A is n^2.
   (c) The total number of lines in A is $n(n + 1)$.

   (Hint: use Major Exercises 4 and 7.)

9. *The real affine plane* has as its "points" all ordered pairs (x, y) of real numbers. A "line" is determined by an ordered triple (u, v, w) of real numbers such that either $u \neq 0$ or $v \neq 0$, and it is defined as the set of all "points" (x, y) satisfying the linear equation $ux + vy + w = 0$. "Incidence" is defined as set membership. Verify that all the axioms for an affine plane are satisfied by this interpretation.

10. A "point" $[x, y, z]$ in *the real projective plane* is determined by an ordered triple (x, y, z) of real numbers that are not all zero, and it consists of all the ordered triples of the form (kx, ky, kz) for all real numbers $k \neq 0$; thus, $[kx, ky, kz] = [x, y, z]$. A "line" in the real projective plane is determined by an ordered triple (u, v, w) of real numbers that are not all zero, and it is defined as the set of all "points" $[x, y, z]$ whose coordinates satisfy the linear equation $ux + vy + wz = 0$. "Incidence" is defined as set membership. Verify that all the axioms for a projective plane are satisfied by this interpretation. Prove that by taking $z = 0$ as the equation of the "line at infinity," by assigning the affine "point" (x, y) the homogeneous coordinates $[x, y, 1]$, and by assigning affine "lines" to projective "lines" in the obvious way, the real projective plane becomes isomorphic to the projective completion of the real affine plane (see Major Exercise 4). Prove that the models in Exercise 10d are also isomorphic to the real projective plane.

11. The following statement is by the French mathematician Desargues: "If the vertices of two triangles correspond in such a way that the lines joining corresponding vertices are concurrent, then the intersections of corresponding sides are collinear." (See Figure 2.10.) This statement is independent of the axioms for projective planes: it holds in the real projective plane, but there exists other projective planes in which it fails. Report on this independence result (see Artzy or Stevenson).

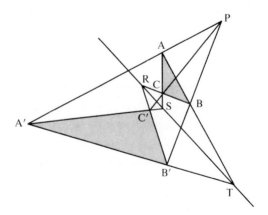

Figure 2.10
Desargues' theorem.

Hilbert's Axioms | 3

The value of Euclid's work as a masterpiece of logic has been very grossly exaggerated.

Bertrand Russell

Flaws in Euclid

Having clarified our rules of reasoning (Chapter 2), let us return to the postulates of Euclid. In Exercises 9 and 10 of Chapter 1 we saw that Euclid neglected to state his assumptions that points and lines exist, that not all points are collinear, and that every line has at least two points lying on it. We made these assumptions explicit in Chapter 2 (p. 42) by adding two more axioms of incidence.

In Exercises 6 and 7, Chapter 1, we saw that some assumptions about "betweenness" are needed. In fact, Euclid never mentioned this notion explicitly, but tacitly assumed certain facts about it that are obvious in diagrams. In Chapter 2 we saw the danger of reasoning from diagrams, so these tacit assumptions will have to be made explicit.

Quite a few of Euclid's proofs are based on reasoning from diagrams. To make these proofs rigorous, a much larger system of explicit axioms is needed. Many such axiom systems have been proposed. We will present a modified version of David Hilbert's system of axioms. Hilbert's system was not the first, but his axioms are perhaps the most intuitive and are certainly the closest in spirit to Euclid's.*

---

* Let us not forget that no serious work toward constructing new axioms for Euclidean geometry had been done until the discovery of non-Euclidean geometry shocked mathematicians into reexamining the foundations of the former. We have the paradox of non-Euclidean geometry helping us to better understand Euclidean geometry!

David Hilbert

During the first quarter of the twentieth century Hilbert was considered the leading mathematician of the world.* He made outstanding, original contributions to a wide range of mathematical fields as well as to physics. He is perhaps best-known for his research in the foundations of geometry as well as the foundations of algebraic number theory, infinite dimensional spaces, and mathematical logic. A great champion of the axiomatic method, he axiomatized all of the above subjects except for physics (although he did succeed in providing physicists with very valuable mathematical techniques). He was also a mathematical prophet; in 1900 he predicted 23 of the most important mathematical problems of this century.

He has been quoted as saying: "One must be able to say at all times—instead of points, lines and planes—tables, chairs and beer mugs." In other words, since no properties of points, lines, and planes may be used in a proof other than the properties given by the axioms, you may as well call these undefined entities by other names.

Hilbert's axioms are divided into five groups: incidence, betweenness, congruence, continuity, and parallelism. We have already seen the three axioms of incidence in Chapter 2 (p. 42). In the next sections we will deal successively with the other groups of axioms.

Axioms of Betweenness

To further illustrate the need for axioms of betweenness, consider the following attempted proof of the theorem that base angles of an isosceles triangle are congruent. This is not Euclid's proof, which is flawed in other ways (see Golos, p. 57), but is an argument found in some high school geometry texts.

PROOF:
Given $\triangle ABC$ with $AC \cong BC$.
To prove $\angle A \cong \angle B$ (see Figure 3.1).
(1) Let the bisector of $\angle C$ meet AB at D (every angle has a bisector).
(2) In triangles $\triangle ACD$ and $\triangle BCD$, $AC \cong BC$ (hypothesis).
(3) $\angle ACD \cong \angle BCD$ (definition of bisector of an angle).
(4) $CD \cong CD$ (things that are equal are congruent).

---

* I heartily recommend the warm and colorful biography of Hilbert by Constance Reid. It is nontechnical and conveys the excitement of the time when Göttingen was the capital of the mathematical world.

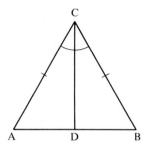

Figure 3.1

(5) $\triangle ACD \cong \triangle BCD$ (SAS).

(6) Therefore, $\sphericalangle A \cong \sphericalangle B$ (corresponding angles of congruent triangles). ∎

Consider the first step, whose justification is that every angle has a bisector. This is a correct statement and can be proved separately. But how do we know that the bisector of $\sphericalangle C$ meets $\overleftrightarrow{AB}$, or if it does, how do we know that the point of intersection D lies *between* A and B? This may seem obvious, but if we are to be rigorous, it requires proof. For all we know, the picture might look like this:

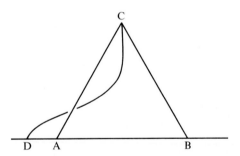

Figure 3.2

If this were the case, Steps 2–5 would still be correct, but we could conclude only that $\sphericalangle B$ is congruent to $\sphericalangle CAD$, not to $\sphericalangle CAB$, since $\sphericalangle CAD$ is the angle in $\triangle ACD$ that corresponds to $\sphericalangle B$.

Once we state our axioms of betweenness, it will be possible to prove (after a considerable amount of work) that the bisector of $\sphericalangle C$ does meet $\overleftrightarrow{AB}$ in a point D between A and B, so the above argument will be repaired (see crossbar theorem, p. 69). There is, however, an easier proof of the theorem (given in the next section). We will use the shorthand notation

$$A * B * C$$

to abbreviate the statement "point B is between point A and point C."

BETWEENNESS AXIOM 1.   If $A * B * C$, then A, B, and C are three distinct points all lying on the same line, and $C * B * A$.

The first part of this axiom fills the gap mentioned in Exercise 6, Chapter 1. The second part $(C * B * A)$ makes the obvious remark that "between A and C" means the same as "between C and A"—it doesn't matter whether A or C is mentioned first.

BETWEENNESS AXIOM 2.   Given any two distinct points B and D, there exist points A, C, and E lying on $\overleftrightarrow{BD}$ such that $A * B * D$, $B * C * D$, and $B * D * E$.

This axiom insures that there are points between B and D and that the line $\overleftrightarrow{BD}$ does not end at either B or D.

A   B   C   D   E   Figure 3.3

BETWEENNESS AXIOM 3.   If A, B, and C are three distinct points lying on the same line, then one and only one of the points is between the other two.

This axiom insures that a line is not circular; if the points were on a circle, you would then have to say that each is between the other two (or none is between the other two—it would depend on which of the two arcs you look at—Figure 3.4).

Before stating the last betweenness axiom, let us examine some consequences of the first three. Recall that the *segment* AB is defined as the set of all points between A and B together with the endpoints A

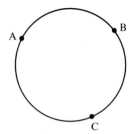

Figure 3.4

and B. The *ray* $\overrightarrow{AB}$ is defined as the set of all points on the segment AB together with all points C such that $A * B * C$. The second axiom insures that such points C exist, so the ray $\overrightarrow{AB}$ is larger than the segment AB. We can now prove the formulas you encountered in Exercise 7, Chapter 1.

PROPOSITION 3.1.   For any two points A and B: (i) $\overrightarrow{AB} \cap \overrightarrow{BA} = AB$, and (ii) $\overrightarrow{AB} \cup \overrightarrow{BA} = \overleftrightarrow{AB}$.

PROOF OF (i):
(1) By definition of segment and ray, $AB \subset \overrightarrow{AB}$ and $AB \subset \overrightarrow{BA}$, so by definition of intersection, $AB \subset \overrightarrow{AB} \cap \overrightarrow{BA}$.
(2) Conversely, let the point C belong to the intersection of $\overrightarrow{AB}$ and $\overrightarrow{BA}$; we wish to show that C belongs to AB.
(3) If $C = A$ or $C = B$, C is an endpoint of AB. Otherwise, A, B, and C are three collinear points (by definition of ray and Axiom 1), so exactly one of the relations $A * C * B$, $A * B * C$, or $C * A * B$ holds (Axiom 3).
(4) If $A * B * C$ holds, then C is not on $\overrightarrow{BA}$; if $C * A * B$ holds, then C is not on $\overrightarrow{AB}$. In either case, C does not belong to both rays.
(5) Hence, the relation $A * C * B$ must hold, so C belongs to AB. ∎

The proof of (ii) is similar and is left as an exercise. (To be truly precise, we should say that $\overrightarrow{AB} \cup \overrightarrow{BA}$ equals the set of points lying on the line $\overleftrightarrow{AB}$, not that it equals the line itself—because "line" is undefined, we do not know that a line is a set of points.)

Recall next that if $C * A * B$, then $\overrightarrow{AC}$ is said to be *opposite* to $\overrightarrow{AB}$:

Figure 3.5

By Axiom 1, points A, B, and C are collinear, and by Axiom 3, C does not belong to $\overleftrightarrow{AB}$, so rays $\overrightarrow{AB}$ and $\overrightarrow{AC}$ are distinct. This definition is therefore in agreement with the definition given in Chapter 1. Axiom 2 guarantees that every ray $\overrightarrow{AB}$ has an opposite ray $\overrightarrow{AC}$.

It seems clear from the above figure that every point P lying on the line l through A, B, C must either belong to ray $\overrightarrow{AB}$ or to the opposite ray $\overrightarrow{AC}$. This statement seems similar to the second assertion of Proposition 3.1, but it is actually more complicated; we are now discussing *four* points A, B, C, and P, whereas previously we had to deal with only three points at a time. In fact, we encounter here another "pictorially obvious" assertion that cannot be proved without introducing another axiom (see Exercise 17).

Suppose we call the assertion "C ∗ A ∗ B and P collinear with A, B, C ⇒ P ε $\overrightarrow{AC}$ ∪ $\overrightarrow{AB}$" *the line separation property.* Some mathematicians take this property as another axiom (e.g., Golos). However, it is considered inelegant in mathematics to assume more axioms than are necessary (although we pay for elegance by having to work harder to prove results that appear obvious). So we will not assume the line separation property as an axiom; instead, we will prove it as a consequence of our previous axioms and our last betweenness axiom, called *the plane separation axiom.*

DEFINITION.   Let l be any line, A and B any points that do not lie on l. If A = B or if segment AB contains no point lying on l, we say A and B are *on the same side of l,* whereas if A ≠ B and segment AB does intersect l, we say that A and B are *on opposite sides* of l [see Figure 3.6]. The law of the excluded middle (Rule 10) tells us that A and B are either on the same side or on opposite sides of l.

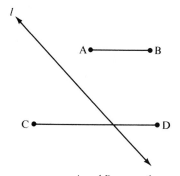

Figure 3.6

A and B are on the same side of l; C and D are on opposite sides of l.

BETWEENNESS AXIOM 4 (Separation). For every line l and for any three points A, B, and C not lying on l:

    (i) if A and B are on the same side of l and B and C are on the same side of l, then A and C are on the same side of l [Figure 3.7].

    (ii) if A and B are on opposite sides of l and B and C are on opposite sides of l, then A and C are on the same side of l [Figure 3.8].

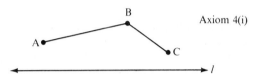

Axiom 4(i)

Figure 3.7

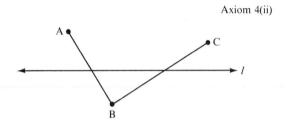

Axiom 4(ii)

Figure 3.8

    Axiom 4(i) indirectly guarantees that our geometry is two-dimensional, since it would not necessarily hold in three-space. (Line l could be outside the plane of this page and cut through segment AC; this interpretation shows that if we assumed the line separation property as an axiom, we could not prove the plane separation property.) Betweenness Axiom 4 is also needed to make sense out of Euclid's fifth postulate, which talks about two lines meeting on one "side" of a transversal. We can now define a *side* of a line l as the set of all points that are on the same side of l as some particular point not lying on l. (This definition may seem circular because we use the word "side" twice, but it is not; we have already defined the compound expression "on the same side.") Another expression commonly used for a "side of l" is a *half-plane bounded by l*.

PROPOSITION 3.2.   Every line bounds exactly two half-planes and these half-planes have no point in common.

PROOF :

(1) For every line *l* there exists a point A not lying on *l* (by Incidence Axiom 3).

(2) There exists a point O lying on *l* (by Incidence Axiom 2).

(3) There exists a point B such that B * O * A (Betweenness Axiom 2).

(4) Then A and B are on opposite sides of *l* (by definition), so *l* has at least two sides.

(5) Let C be any point distinct from A and B and not lying on *l*. If C and B are not on the same side of *l*, then C and A are on the same side of *l* (by the law of excluded middle and Betweenness Axiom 4(ii)).

(6) Hence, *l* has exactly two sides (by Steps 4 and 5).

(7) If these sides had a point C in common (RAA hypothesis), then by Betweenness Axiom 4(i) A and B would be on the same side of *l*, contradiction. ∎

We next apply the plane separation property to study betweenness relations among four points.

PROPOSITION 3.3.   Given A * B * C and A * C * D. Then B * C * D and A * B * D. [See Figure 3.9.]

PROOF :

(1) A, B, C, and D are four distinct collinear points (see Exercise 1).

(2) There exists a point E not on the line through A, B, C, D (Incidence Axiom 3).

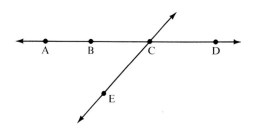

Figure 3.9

(3) Consider line $\overleftrightarrow{EC}$. Since (by hypothesis) AD meets this line in point C, A and D are on opposite sides of $\overleftrightarrow{EC}$.

(4) We claim A and B are on the same side of $\overleftrightarrow{EC}$. Assume on the contrary that A and B are on opposite sides of $\overleftrightarrow{EC}$ (RAA hypothesis).

(5) Then $\overleftrightarrow{EC}$ meets $\overrightarrow{AB}$ in a point between A and B (definition of "opposite sides").

(6) That point must be C (Proposition 2.1).

(7) Thus, A * B * C and A * C * B, which contradicts Betweenness Axiom 3.

(8) Hence, A and B are on the same side of $\overleftrightarrow{EC}$ (RAA conclusion).

(9) It follows from Steps 3 and 8 and Proposition 3.2 that B and D must be on opposite sides of $\overleftrightarrow{EC}$.

(10) Hence, the point C of intersection of lines $\overleftrightarrow{EC}$ and $\overleftrightarrow{BD}$ lies between B and D (definition of "opposite sides"; Proposition 2.1, i.e., that the point of intersection is unique).

A similar argument involving $\overleftrightarrow{EB}$ proves that A * B * D (Exercise 2b). ■

Finally we prove the *line separation property*.

PROPOSITION 3.4.   If C * A * B and l is the line through A, B, and C (Betweenness Axiom 1), then for every point P lying on l, P lies either on ray $\overrightarrow{AB}$ or on the opposite ray $\overrightarrow{AC}$.

PROOF:

(1) Either P lies on $\overrightarrow{AB}$ or it does not (law of excluded middle).

(2) If P does lie on $\overrightarrow{AB}$, we are done, so assume it doesn't; then P * A * B (Betweenness Axiom 3).

(3) If P = C then P lies on $\overrightarrow{AC}$ (by definition), so assume P ≠ C; then exactly one of the relations C * A * P, C * P * A, or P * C * A holds (Betweenness Axiom 3 again).

(4) Suppose the relation C * A * P holds (RAA hypothesis).

(5) We know (by Betweenness Axiom 3) that exactly one of the relations P * C * B, C * P * B, or C * B * P holds.

(6) If P * B * C, then combining this with P * A * B (Step 2) gives A * B * C (Proposition 3.3), contradicting the hypothesis.

(7) If C * P * B, then combining this with C * A * P (Step 4) gives A * P * B (Proposition 3.3), contradicting Step 2.

(8) If B * C * P, then combining this with B * A * C (hypothesis and

Betweenness Axiom 1) gives A * C * P (Proposition 3.3), contradicting Step 4.

(9) Since we obtain a contradiction in all three cases, C * A * P does not hold (RAA conclusion).

(10) Therefore, C * P * A or P * C * A (Step 3), which means that P lies on the opposite ray $\overrightarrow{AC}$.∎

The next theorem states a visually obvious property that Pasch discovered Euclid to be using without proof.

PASCH'S THEOREM.   If △ABC is any triangle and *l* is any line intersecting side AB in a point between A and B, then *l* also intersects either side AC or side BC [see Figure 3.10]. If C does not lie on *l*, then *l* does not intersect both AC and BC.

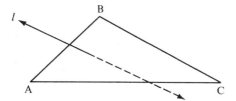

Figure 3.10

Intuitively, this theorem says that if a line "goes into" a triangle through one side, it must "come out" through another side.

PROOF:

(1) Either C lies on *l* or it does not; if it does, the theorem holds (law of excluded middle).

(2) A and B do not lie on *l*, and the segment AB does intersect *l* (hypothesis).

(3) Hence, A and B lie on opposite sides of *l* (by definition).

(4) From Step 1 we may assume that C does not lie on *l*, in which case C is either on the same side of *l* as A or on the same side of *l* as B (separation axiom).

(5) If C is on the same side of *l* as A, then C is on the opposite side from B, which means that *l* intersects BC and does not intersect AC; similarly if C is on the same side of *l* as B, then *l* intersects AC and does not intersect BC (separation axiom).∎

Here are some more results on betweenness and separation that you will be asked to prove in the exercises.

PROPOSITION 3.5.   Given A * B * C. Then AC = AB ∪ BC and B is the only point common to segments AB and BC.

PROPOSITION 3.6.   Given A * B * C. Then B is the only point common to rays $\overrightarrow{BA}$ and $\overrightarrow{BC}$, and $\overrightarrow{AB} = \overrightarrow{AC}$.

DEFINITION.   Given an angle ⊰ CAB, define a point D to be in the *interior* of ⊰ CAB if D is on the same side of $\overleftrightarrow{AC}$ as B and if D is also on the same side of $\overleftrightarrow{AB}$ as C. (Thus, the interior of an angle is the intersection of two half-planes.)

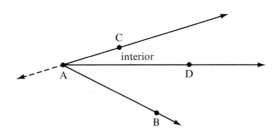

Figure 3.11

PROPOSITION 3.7.   Given an angle ⊰ CAB and point D lying on line $\overleftrightarrow{BC}$. Then D is in the interior of ⊰ CAB if and only if B * D * C [see Figure 3.12].

Warning! Do not assume that every point in the interior of an angle lies on a segment joining a point on one side of the angle to a point on the other side. In fact, this assumption is false in hyperbolic geometry.

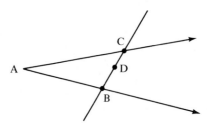

Figure 3.12

PROPOSITION 3.8.   If D is in the interior of ⊰ CAB, then (a) so
is every other point on ray A⃗D except A ; (b) no point on the opposite
ray to A⃗D is in the interior of ⊰ CAB; (c) if C * A * E, then B is in the
interior of ⊰ DAE [see Figure 3.13].

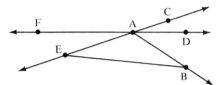

Figure 3.13

DEFINITION.   Ray A⃗D is *between* rays A⃗C and A⃗B if A⃗B and A⃗C
are not opposite rays and D is interior to ⊰ CAB. (By Proposition
3.8a, this definition does not depend on the choice of point D on A⃗D.)

CROSSBAR THEOREM.   If A⃗D is between A⃗C and A⃗B, then A⃗D
intersects segment BC [see Figure 3.14].

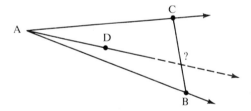

Figure 3.14

DEFINITION.   The *interior* of a triangle is the intersection of the
interiors of its three angles. Define a point to be *exterior* to the triangle
if it is not in the interior and does not lie on any side of the triangle.

PROPOSITION 3.9.   (a) If a ray *r* emanating from an exterior point
of △ABC intersects side AB in a point between A and B, then *r* also
intersects side AC or side BC. (b) If a ray emanates from an interior
point of △ABC, then it intersects one of the sides, and if it does not
pass through a vertex, it intersects only one side.

Axioms of Congruence

Recall that "congruent" is the last of our undefined terms; it is either a relation between segments or a relation between angles. We are accustomed to congruence as a relation between triangles, but we can now define this as follows: Two triangles are *congruent* if a one-to-one correspondence can be set up between their vertices so that corresponding sides are congruent and corresponding angles are congruent. When we write $\triangle ABC \cong \triangle DEF$ we understand that A corresponds to D, B to E, and C to F. Similar definitions can be given for congruence of quadrilaterals, pentagons, etc.

CONGRUENCE AXIOM 1.   If A and B are distinct points and if A′ is any point, then for each ray r emanating from A′ there is a *unique* point B′ on r such that B′ ≠ A′ and AB ≅ A′B′.

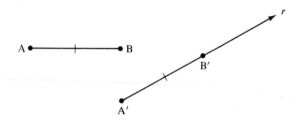

Figure 3.15

Intuitively speaking, this axiom says you can "move" the segment AB so that it lies on the ray r with A superimposed on A′, B superimposed on B′. (In Major Exercise 2, Chapter 1, you showed how to do this with a straightedge and collapsible compass.)

CONGRUENCE AXIOM 2.   If AB ≅ CD and AB ≅ EF, then CD ≅ EF. Moreover, every segment is congruent to itself.

This axiom replaces Euclid's first Common Notion, since it says that segments congruent to the same segment are congruent to each other. It also replaces the fourth Common Notion, since it says that segments that coincide are congruent.

CONGRUENCE AXIOM 3.   If A * B * C, A′ * B′ * C′, AB ≅ A′B′, and BC ≅ B′C′, then AC ≅ A′C′.

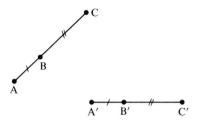

Figure 3.16

This axiom replaces the second Common Notion, since it says that if congruent segments are "added" to congruent segments, the sums are congruent. Here, "adding" means juxtaposing segments along the same line. For example, using Congruence Axioms 1 and 3, you can lay off a copy of a given segment AB two, three, ..., n times, to get a new segment $n \cdot AB$.

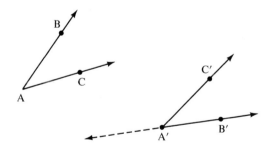

Figure 3.17

$AB'' = 3 \cdot AB$.

CONGRUENCE AXIOM 4.   Given any angle ⊰ BAC (where, by definition of "angle," $\overrightarrow{AB}$ is not opposite to $\overrightarrow{AC}$), and given any ray $\overrightarrow{A'B'}$ emanating from a point A', then there is a *unique* ray $\overrightarrow{A'C'}$ on a given side of line $\overleftrightarrow{A'B'}$ such that ⊰ B'A'C' ≅ ⊰ BAC.

Figure 3.18

This axiom can be paraphrased to state that a given angle can be "laid off" on a given side of a given ray in a unique way (see Major Exercise 1g, Chapter 1).

CONGRUENCE AXIOM 5.   If ⊰ A ≅ ⊰ B and ⊰ A ≅ ⊰ C, then ⊰ B ≅ ⊰ C. Moreover, every angle is congruent to itself.

This is the analog for angles of Congruence Axiom 2 for segments; the first part asserts the transitivity and the second part the reflexivity of the congruence relation. Combining them, we can prove the symmetry of this relation: ⊁ A ≅ ⊁ B ⇒ ⊁ B ≅ ⊁ A.

PROOF: ⊁ A ≅ ⊁ B (hypothesis) and ⊁ A ≅ ⊁ A (reflexivity) imply (substituting A for C in Congruence Axiom 5) ⊁ B ≅ ⊁ A (transitivity). ∎

It would seem natural to assume next an "addition axiom" for congruence of *angles* analogous to Congruence Axiom 3 (the addition axiom for congruence of segments). We won't do this, however, because such a result can be proved using the next congruence axiom (see Proposition 3.19).

CONGRUENCE AXIOM 6 (SAS). If two sides and the included angle of one triangle are congruent respectively to two sides and the included angle of another triangle, then the two triangles are congruent.

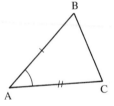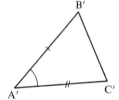

Figure 3.19

This is the side-angle-side criterion for congruence of triangles with which you are familiar. As mentioned, Euclid attempted to prove SAS as a theorem. His argument was essentially as follows: Move △A'B'C' so as to place point A' on point A and $\overleftrightarrow{A'B'}$ on $\overleftrightarrow{AB}$. Since AB ≅ A'B', by hypothesis, point B' must fall on point B. Since ⊁ A ≅ ⊁ A', $\overleftrightarrow{A'C'}$ must fall on $\overleftrightarrow{AC}$, and since AC ≅ A'C', point C' must coincide with point C. Hence, B'C' will coincide with BC and the remaining angles will coincide with the remaining angles, so the triangles will be congruent.

This argument is called *superposition*. It derives from the experience of drawing two triangles on paper, cutting out one, and placing it on top of the other. Although this is a good way to convince a novice in geometry to accept SAS, it is not a proof, and Euclid reluctantly used it in only one other theorem. It is not a proof because Euclid never stated an axiom that allows figures to be moved around without changing their size and shape.

Some modern writers introduce "motion" as an undefined term and lay down axioms for this term. (In fact, in Pieri's foundations of geometry, "point" and "motion" are the only undefined terms.) Or else, the geometry is first built up on a different basis, "distances" introduced, and a "motion" defined as a one-to-one transformation of the plane onto itself that preserves distance. Euclid can be vindicated by either approach. In fact, Felix Klein, in his 1872 Erlanger *Programm,* defined geometry as the study of those properties of figures that remain invariant under a particular group of transformations. (For an approach to geometry using motion as a foundation, see Ewald; see also the Major Exercises in Chapter 7.)

The "motions" mentioned in the previous paragraph are all mathematical abstractions. The physicist Baron von Helmholtz took a different approach. He maintained that geometry requires us to assume the actual existence of rigid bodies with free mobility in space. Geometry is then dependent on mechanics. Naturally, this viewpoint was rejected by most mathematicians.

As an application of SAS, the simple proof of Pappus (300 A.D.) for the theorem on base angles of an isosceles triangle follows.

PROPOSITION 3.10. If in $\triangle ABC$ we have $AB \cong AC$, then $\angle B \cong \angle C$.

PROOF:

(1) Consider the correspondence of vertices $A \leftrightarrow A$, $B \leftrightarrow C$, $C \leftrightarrow B$. Under this correspondence, two sides and the included angle of $\triangle ABC$ are congruent respectively to the corresponding sides and included

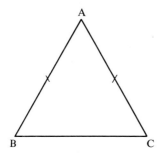

Figure 3.20

angle of △ACB (by hypothesis and Congruence Axiom 5 that an angle is congruent to itself).

(2) Hence, △ABC ≅ △ACB (SAS), so the corresponding angles ⊰ B and ⊰ C are congruent (by definition of congruence of triangles). ■

Here are some more familiar results on congruence. We will prove some of them; if the proof is omitted, see the exercises.

PROPOSITION 3.11 (Segment Subtraction).   If A ∗ B ∗ C, D ∗ E ∗ F, AB ≅ DE, and AC ≅ DF, then BC ≅ EF.

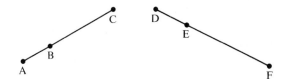

Figure 3.21

PROPOSITION 3.12.   Given AC ≅ DF, then for any point B between A and C, there is a unique point E between D and F such that AB ≅ DE.

PROOF:

(1) There is a unique point E on D$\overrightarrow{F}$ such that AB ≅ DE (Congruence Axiom 1).

(2) Suppose E were not between D and F (RAA hypothesis).

(3) Then either E = F or D ∗ F ∗ E (definition of D$\overrightarrow{F}$).

(4) If E = F, then B and C are two distinct points on A$\overrightarrow{C}$ such that AC ≅ DF ≅ AB (hypothesis, Step 1), contradicting the uniqueness part of Congruence Axiom 1.

(5) If D ∗ F ∗ E, then there is a point G on the ray opposite to C$\overrightarrow{A}$ such that FE ≅ CG (Congruence Axiom 1).

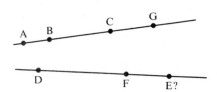

Figure 3.22

(6) Then AG ≅ DE (Congruence Axiom 3).

(7) Thus, there are two distinct points B and G on $\overrightarrow{AC}$ such that AG ≅ DE ≅ AB (Steps 1, 5, and 6), contradicting the uniqueness part of Congruence Axiom 1.

(8) D * E * F (RAA conclusion). ∎

DEFINITION.   AB < CD (or CD > AB) means that there exists a point E between C and D such that AB ≅ CE.

PROPOSITION 3.13 (Segment Ordering).   (a) Exactly one of the following conditions holds (trichotomy): AB < CD, AB ≅ CD, or AB > CD. (b) If AB < CD and CD ≅ EF, then AB < EF. (c) If AB > CD and CD ≅ EF, then AB > EF. (d) If AB < CD and CD < EF, then AB < EF (transitivity).

PROPOSITION 3.14   Supplements of congruent angles are congruent.

PROPOSITION 3.15.   (a) Vertical angles are congruent to each other. (b) An angle congruent to a right angle is a right angle.

PROPOSITION 3.16.   For every line *l* and every point P there exists a line through P perpendicular to *l*.

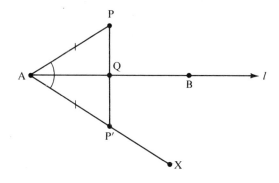

Figure 3.23

PROOF:

(1) Assume first that P does not lie on *l* and let A and B be any two points on *l* (Incidence Axiom 2).

(2) On the opposite side of l from P there exists a ray $\overrightarrow{AX}$ such that $\sphericalangle\,XAB \cong \sphericalangle\,PAB$ (Congruence Axiom 4).

(3) There is a point P′ on $\overrightarrow{AX}$ such that AP′ $\cong$ AP (Congruence Axiom 1).

(4) PP′ intersects l in a point Q (definition of opposite sides of l).

(5) If Q = A, then $\overleftrightarrow{PP'} \perp l$ (definition of $\perp$).

(6) If Q $\neq$ A, then $\triangle$PAQ $\cong$ $\triangle$P′AQ (SAS).

(7) Hence, $\sphericalangle\,$PQA $\cong$ $\sphericalangle\,$P′QA (corresponding angles), so $\overleftrightarrow{PP'} \perp l$ (definition of $\perp$).

(8) Assume now that P lies on l. Since there are points not lying on l (Incidence Axiom 3), we can drop a perpendicular from one of them to l (Steps 5 and 7), thereby obtaining a right angle.

(9) We can lay off an angle congruent to this right angle with vertex at P and one side on l (Congruence Axiom 4); the other side of this angle is part of a line through P perpendicular to l (Proposition 3.15b). ∎

It is natural to ask whether the perpendicular to l through P constructed in Proposition 3.16 is unique. If P lies on l, the uniqueness part of Congruence Axiom 4 guarantees that the perpendicular is unique. If P does not lie on l, we will not be able to prove uniqueness for the perpendicular until the next chapter. In elliptic geometry there is a point P called the "pole" of l such that every line through P is perpendicular to l! To visualize this, think of l as the equator on a sphere and P as the north pole; every great circle through the north pole is perpendicular to the equator (Figure 3.24).

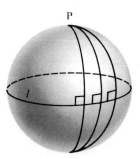

Figure 3.24

PROPOSITION 3.17 (ASA Criterion for Congruence).   Given $\triangle$ABC and $\triangle$DEF with $\sphericalangle\,$A $\cong$ $\sphericalangle\,$D, $\sphericalangle\,$C $\cong$ $\sphericalangle\,$F, and AC $\cong$ DF. Then $\triangle$ABC $\cong$ $\triangle$DEF.

PROPOSITION 3.18 (Converse of Proposition 3.10).   If in $\triangle ABC$ we have $\measuredangle\, B \cong \measuredangle\, C$, then $AB \cong AC$ and $\triangle ABC$ is isosceles.

PROPOSITION 3.19 (Angle Addition).   Given $\overrightarrow{BG}$ between $\overrightarrow{BA}$ and $\overrightarrow{BC}$, $\overrightarrow{EH}$ between $\overrightarrow{ED}$ and $\overrightarrow{EF}$, $\measuredangle\, CBG \cong \measuredangle\, FEH$, and $\measuredangle\, GBA \cong \measuredangle\, HED$. Then $\measuredangle\, ABC \cong \measuredangle\, DEF$.

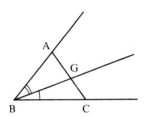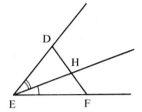

Figure 3.25

PROOF :
(1) By the crossbar theorem, we assume G is chosen so that $A * G * C$.
(2) By Congruence Axiom 1, we assume D, F, and H chosen so that $AB \cong ED$, $GB \cong EH$, and $CB \cong EF$.
(3) Then $\triangle ABG \cong \triangle DEH$ and $\triangle GBC \cong \triangle HEF$ (SAS).
(4) $\measuredangle\, DHE \cong \measuredangle\, AGB$, $\measuredangle\, FHE = \measuredangle\, CGB$ (Step 3), and $\measuredangle\, AGB$ is supplementary to $\measuredangle\, CGB$ (Step 1).
(5) $\measuredangle\, DHE$ is supplementary to $\measuredangle\, FHE$ (Step 4 and Proposition 3.14) so D, H, and F are collinear.
(6) $D * H * F$ (Proposition 3.7, using the hypothesis on $\overrightarrow{EH}$).
(7) $AC \cong DF$ (Steps 3 and 6, Congruence Axiom 3).
(8) $\triangle ABC \cong \triangle DEF$ (SAS, Steps 2, 3, and 7).
(9) $\measuredangle\, ABC \cong \measuredangle\, DEF$ (corresponding angles). ■

PROPOSITION 3.20 (Angle Subtraction).   Given $\overrightarrow{BG}$ between $\overrightarrow{BA}$ and $\overrightarrow{BC}$, $\overrightarrow{EH}$ between $\overrightarrow{ED}$ and $\overrightarrow{EF}$, $\measuredangle\, CBG \cong \measuredangle\, FEH$, and $\measuredangle\, ABC \cong \measuredangle\, DEF$. Then $\measuredangle\, GBA \cong \measuredangle\, HED$.

DEFINITION. $\measuredangle\, ABC < \measuredangle\, DEF$ means there is a ray $\overrightarrow{EG}$ between $\overrightarrow{ED}$ and $\overrightarrow{EF}$ such that $\measuredangle\, ABC \cong \measuredangle\, GEF$ (see Figure 3.26).

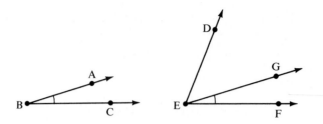

Figure 3.26

PROPOSITION 3.21 (Ordering of Angles). (a) Exactly one of the following conditions holds (trichotomy): ⊰P < ⊰Q, ⊰P ≅ ⊰Q, or ⊰Q < ⊰P. (b) if ⊰P < ⊰Q and ⊰Q ≅ ⊰R, then ⊰P < ⊰R. (c) If ⊰P > ⊰Q and ⊰Q ≅ ⊰R, then ⊰P > ⊰R. (d) If ⊰P < ⊰Q and ⊰Q < ⊰R, then ⊰P < ⊰R.

PROPOSITION 3.22 (SSS Criterion for Congruence). Given △ABC and △DEF. If AB ≅ DE, BC ≅ EF, and AC ≅ DF, then △ABC ≅ △DEF.

The AAS criterion for congruence will be given in the next chapter because its proof is more difficult. The next proposition was assumed as an axiom by Euclid, but can be proved from Hilbert's axioms.

PROPOSITION 3.23 (Euclid's Fourth Postulate). All right angles are congruent to each other.

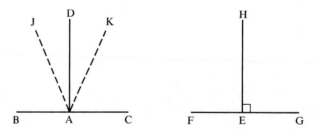

Figure 3.27

PROOF:

(1) Given ⊰ BAD ≅ ⊰ CAD and ⊰ FEH ≅ ⊰ GEH (two pairs of right angles, by definition). Assume the contrary, that ⊰ BAD is not congruent to ⊰ FEH (RAA hypothesis).

(2) Then one of these angles is smaller than the other, e.g., ⦨ FEH < ⦨ BAD (Proposition 3.21a), so that by definition there is a ray $\overrightarrow{AJ}$ between $\overrightarrow{AB}$ and $\overrightarrow{AD}$ such that ⦨ BAJ ≅ ⦨ FEH.

(3) ⦨ CAJ ≅ ⦨ GEH (Proposition 3.14).

(4) ⦨ CAJ ≅ ⦨ FEH (Steps 1 and 3, Congruence Axiom 5).

(5) There is a ray $\overrightarrow{AK}$ between $\overrightarrow{AD}$ and $\overrightarrow{AC}$ such that ⦨ BAJ ≅ ⦨ CAK (Step 1 and Proposition 3.21b).

(6) ⦨ BAJ ≅ ⦨ CAJ (Steps 2 and 4, and Congruence Axiom 5).

(7) ⦨ CAJ ≅ ⦨ CAK (Steps 5 and 6, and Congruence Axiom 5).

(8) Thus, we have ⦨ CAD greater than ⦨ CAK (by definition) and less than its congruent angle ⦨ CAJ (Step 7 and Proposition 3.8c), which contradicts Proposition 3.21.

(9) ⦨ BAD ≅ ⦨ FEH (RAA conclusion). ∎

Axioms of Continuity

These axioms are the most subtle and difficult to comprehend. They are needed to fill in a number of gaps in Euclid's *Elements*. A thorough discussion of them is beyond the scope of this book; it will suffice for our purposes to state them and make only brief remarks.

ARCHIMEDES' AXIOM. If AB and CD are any segments, then there is a number n such that if segment CD is laid off n times on the ray $\overrightarrow{AB}$ emanating from A, then a point E is reached where $n \cdot CD \cong AE$ and B is between A and E.

For example, if AB were π units long and CD of one unit length, you would have to lay off CD at least four times to get to a point E beyond the point B (see Figure 3.28).

The intuitive content of Archimedes' axiom is that if you arbitrarily choose one segment CD as a unit of length, then every other segment has

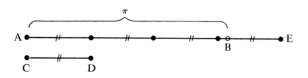

Figure 3.28

finite length with respect to this unit (in the notation of the axiom the length of AB with respect to CD as unit is at most n units). Another way to look at it is to choose AB as unit of length. The axiom says that no other segment can be infinitesimally small with respect to this unit (the length of CD with respect to AB as unit is at least $1/n$ units).

DEDEKIND'S AXIOM.*   Suppose that the set of all points on a line l is the union $\Sigma_1 \cup \Sigma_2$ of two nonempty subsets such that no point of Σ_1 is between two points of Σ_2 and vice versa. Then there is a unique point O lying on l such that $P_1 * O * P_2$ if and only if $P_1 \varepsilon \Sigma_1$ and $P_2 \varepsilon \Sigma_2$ and $O \neq P_1, P_2$.

Figure 3.29

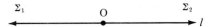

Dedekind's axiom is a sort of converse to the line separation property (p. 63). That property says that any point O on l separates all the other points on l into those to the left of O and those to the right. Dedekind's axiom says that conversely, any separation of points on l into left and right is produced by a unique point O. A pair of subsets Σ_1 and Σ_2 with the properties in Dedekind's axiom is called a *Dedekind cut* of the line.

Of course, we must define what is meant by "left" and "right" here. The intuitive idea is obvious; it is left to you to write down a precise definition (Exercise 7).

Loosely speaking, the purpose of Dedekind's axiom is to insure that a line l has no "holes" in it, in the sense that for any point O on l and any positive real number x there exist unique points P_{-x} and P_x on l such that $P_{-x} * O * P_x$ and segments $P_{-x}O$ and OP_x both have length x (with respect to some unit segment of measurement).

Figure 3.30

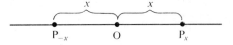

Without Dedekind's axiom there would be no guarantee, for example, of the existence of a segment of length π. With it, we can introduce a rectangular coordinate system into the plane and do geometry analytically, as Descartes and Fermat discovered in the seventeenth century. This

---

* This axiom was proposed by J. W. R. Dedekind in 1871; an analog of it is used in analysis texts to express the completeness of the real number system.

coordinate system enables us to prove that our axioms for Euclidean geometry are *categorical* in the sense that the system has a unique model (up to isomorphism—see Exercise 10, Chapter 2), namely, the usual Cartesian coordinate plane of all ordered pairs of real numbers.

If we omitted Dedekind's axiom, then another model would be the so-called *surd plane,* a plane that is used to prove the impossibility of trisecting every angle with a straightedge and compass (see Moise, p. 228 ff.). The categorical natural of all the axioms is proved in Borsuk-Szmielew (p. 276 ff.).

It can actually be proved that Archimedes' axiom is a consequence of Dedekind's and the other axioms (see Major Exercise 1 or Borsuk-Szmielew, p. 154). To see why Dedekind's axiom is needed, consider the argument Euclid gives to justify his very first proposition.

EUCLID'S PROPOSITION 1.   Given any segment, there is an equilateral triangle having the given segment as one of its sides.

EUCLID'S PROOF:
(1) Let AB be the given segment. With center A and radius AB, let the circle BCD be described (Postulate III).
(2) Again with center B and radius BA, let the circle ACE be described (Postulate III).
(3) From a point C in which the circles cut one another, draw the segments CA and CB (Postulate I).
(4) Since A is the center of the circle CDB, AC is congruent to AB (definition of circle).
(5) Again, since B is the center of circle CAE, BC is congruent to BA (definition of circle).
(6) Since CA and CB are each congruent to AB (Steps 4 and 5), they are congruent to each other (first Common Notion).
(7) Hence, △ABC is an equilateral triangle (by definition) having AB as one of its sides. ■

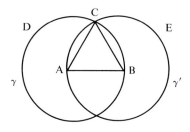

Figure 3.31

Since every step has apparently been justified, you may not see the gap in the proof. It occurs in the first three steps, especially in the third step, which explicitly states that C is a point in which the circles cut each other. (The second step states this implicitly by using the same letter "C" to denote part of the circle, as in the first step.) The point is: how do we know that such a point C exists?

If you believe it is obvious from the diagram that such a point C exists, you are right—but you are not allowed to use the diagram to justify this! We aren't saying that the circles constructed do not cut each other; we're saying only that another axiom is needed to *prove* that they do.

The gap can be filled by proving the following *circular continuity principle*:

CIRCULAR CONTINUITY PRINCIPLE.   If a circle γ has one point inside and one point outside another circle γ', then the two circles intersect in two points.

Here a point P is defined as *inside* a circle with center O and radius OR if OP < OR (*outside* if OP > OR). In Figure 3.31, point B is inside circle γ', and the point B' (not shown) such that A is the midpoint of BB' is outside γ'. The circular continuity principle is a consequence of Dedekind's axiom (see Heath, p. 238).

Another gap in the *Elements* occurs in Euclid's method for dropping a perpendicular to a line (his twelfth proposition). In his construction Euclid tacitly assumes the following *elementary continuity principle*:

ELEMENTARY CONTINUITY PRINCIPLE.   If one endpoint of a segment is inside a circle and the other outside, then the segment intersects the circle.

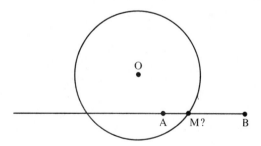

Figure 3.32

Let us sketch a proof of this proposition using Dedekind's axiom:

By definition of "inside" and "outside" of a circle, if O is the center of the circle and OR a radius, we have OA < OR < OB. We define a Dedekind cut for the line $\overleftrightarrow{AB}$ as follows: let Σ_1 be the set of all points P of segment AB such that OP < OR, together with all points on the ray opposite to $\overrightarrow{AB}$; let Σ_2 be the set of all points P of AB such that OP $\cong$ OR or OP > OR, together with all points on the ray opposite to $\overrightarrow{BA}$. If $\overleftrightarrow{AB}$ is directed so that A is to the left of B, then every point of Σ_1 is to the left of every point of Σ_2. Hence, by Dedekind's axiom, there is a unique point M on $\overleftrightarrow{AB}$ such that $P_1 * M * P_2$ for all $P_1 \, \varepsilon \, \Sigma_1$ and $P_2 \, \varepsilon \, \Sigma_2$. We can then show by an RAA argument that OM $\cong$ OR, i.e., M lies on the circle (see Major Exercise 2). ■

It is apparent that such reasoning with Dedekind's axiom is not elementary. If you have difficulty following it, don't be discouraged—it will not play an important role in this book. We will use the continuity axioms only on those few occasions when they prove absolutely essential. On other occasions they can be avoided. For example, Euclid used the existence of equilateral triangles on a given base to prove the existence of angle bisectors and midpoints (see Heath, Propositions 9 and 10), so that the existence of the bisectors and midpoints would seem to depend indirectly on continuity. But there is an ingenious way to prove the existence of angle bisectors and midpoints without using continuity (see Exercises 12–14, Chapter 4).

One reason Dedekind's axiom is not elementary is that it refers to variable sets of points that form Dedekind cuts. Alfred Tarski has proposed a different system of axioms for Euclidean geometry, in which the only variables are points and every statement is made in terms of the undefined term "point" and the undefined relations "betweenness" and "equidistance." With this restricted language, Tarski has been able to express all the results found in traditional elementary geometry textbooks. However, Dedekind's axiom cannot be expressed in this language. Tarski proposes to replace Dedekind's axiom with the elementary continuity principle. He calls this system "the geometry of elementary constructions" because the only existence statements that are valid in the system are those that can be proved by straightedge-and-compass constructions (see his paper "What is elementary geometry?" in Henkin, Suppes and Tarski).

Axiom of Parallelism

If we were to stop with the axioms we now have, we could do quite a bit of geometry, but we still couldn't do all of Euclidean geometry. We would be able to do *absolute geometry,* a misleading name first used by Janos Bolyai and still widely used. I prefer the name suggested by W. Prenowitz and M. Jordan, *neutral geometry,* so called because in doing this geometry we remain neutral about the one axiom from Hilbert's list left to be considered—historically the most controversial axiom of all.

HILBERT'S AXIOM OF PARALLELISM. For every line *l* and every point P not lying on *l* there is at most one line *m* through P such that *m* is parallel to *l*.

Figure 3.33

Note that this axiom is weaker than the Euclidean parallel postulate introduced in Chapter 1 (p. 17). This axiom asserts only that *at most* one line through P is parallel to *l,* whereas the Euclidean parallel postulate asserts in addition that *at least* one line through P is parallel to *l.* The reason "at least" is omitted from Hilbert's axiom is that it can be proved from the other axioms (see Corollary 2 to Theorem 4.1, p. 97); it is therefore unnecessary to assume this as part of an axiom. This observation is important because it implies that the elliptic parallel property (no parallel lines exist) is inconsistent with the axioms of neutral geometry. Thus, a different set of axioms is needed for the foundation of elliptic geometry (see Appendix B).

The axiom of parallelism completes our list of sixteen axioms for Euclidean geometry. In referring to these axioms we will use the following shorthand: the incidence axioms will be denoted by I-1, I-2, and I-3; the betweenness axioms by B-1, B-2, B-3, and B-4; the congruence axioms by C-1, C-2, C-3, C-4, C-5 and C-6. The continuity axioms and the parallelism axiom will be referred to by name.

Review Exercise

Which of the following statements are correct?

(1) Hilbert's axiom of parallelism is the same as the Euclidean parallel postulate given in Chapter 1.

(2) $A * B * C$ is logically equivalent to $C * B * A$.

(3) In Axiom B-2 it is unnecessary to assume the existence of a point E such that $B * D * E$ because this can be proved from the rest of the axiom and Axiom B-1, by interchanging the roles of B and D and taking E to be A.

(4) If A, B, and C are distinct collinear points, it is possible that *both* $A * B * C$ *and* $A * C * B$.

(5) The "line separation property" asserts that a line has two sides.

(6) If points A and B are on opposite sides of a line l, then a point C not on l must either be on the same side of l as A or on the same side of l as B.

(7) If line m is parallel to line l, then all the points on m lie on the same side of l.

(8) If we were to take Pasch's theorem as an axiom instead of the Separation Axiom B-4, then B-4 could be proved as a theorem. Prove it

(9) The notion of "congruence" for two triangles is not defined in this chapter.

(10) It is an immediate consequence of Axiom C-2 that if $AB \cong CD$, then $CD \cong AB$.

(11) One of the congruence axioms asserts that if congruent segments are "subtracted" from congruent segments, the differences are congruent.

(12) In the statement of Axiom C-4 the variables A, B, C, A', and B' are quantified universally, and the variable C' is quantified existentially.

(13) One of the congruence axioms is the side-side-side (SSS) criterion for congruence of triangles.

(14) Euclid attempted unsuccessfully to prove the SAS criterion for congruence by a method called "superposition."

(15) We can use Pappus' method to prove the converse of the theorem on base angles of an isosceles triangle if we first prove the angle-side-angle (ASA) criterion for congruence.

(16) Archimedes' axiom is independent of the other fifteen axioms for Euclidean geometry given in this book.

(17) AB < CD means that there is a point E between C and D such that AB ≅ CE.

(18) Neutral geometry used to be called "absolute geometry"; it is the geometry you have when the axiom of parallelism is excluded from the system of axioms given here.

Exercises on Betweenness

1. Given A * B * C and A * C * D. Prove that:

   (a) A, B, C, D are four distinct points;
   (b) A, B, C, D are collinear (the proof of (a) requires an axiom).

2. (a) Finish the proof of Proposition 3.1 by showing that $\overrightarrow{AB} \cup \overrightarrow{BA} = \overleftrightarrow{AB}$.
   (b) Finish the proof of Proposition 3.3 by showing that A * B * D.
   (c) Prove the converse of Proposition 3.3 by applying the symmetry of betweenness (B-1).

3. Given A * B * C.

   (a) Use Proposition 3.3 to prove that AB ⊂ AC. Interchanging A and C, deduce CB ⊂ CA; which axiom justifies this interchange?
   (b) Use Axiom B-4 to prove that AC ⊂ AB ∪ BC. (Hint: if P is a fourth point on AC, use another line through P to show P ε AB or P ε BC.)
   (c) Finish the proof of Proposition 3.5. (Hint: if P ≠ B and P ε AB ∩ BC, use another line through P to get a contradiction.)

4. Given A * B * C.

   (a) If P is a fourth point collinear with A, B, and C, use Proposition 3.3 and an axiom to prove that ∼A * B * P ⇒ ∼A * C * P.
   (b) Deduce that $\overrightarrow{BA} \subset \overrightarrow{CA}$ and, symmetrically, $\overrightarrow{BC} \subset \overrightarrow{AC}$.
   (c) Use this result, Proposition 3.1a, Proposition 3.3, and Proposition 3.5 to prove that B is the only point that $\overrightarrow{BA}$ and $\overrightarrow{BC}$ have in common.

5. Given A * B * C. Prove that $\overrightarrow{AB} = \overrightarrow{AC}$, completing the proof of Proposition 3.6. Deduce that every ray has a *unique* opposite ray.

6. In Axiom B-2 we were given distinct points B and D and we asserted the existence of points A, C, and E such that A * B * D, B * C * D, and B * D * E. We can now show that it was not necessary to assume the existence of a point C between B and D because we can prove from our other axioms (including the rest of Axiom B-2) and from Pasch's theorem (which was proved without using Axiom B-2) that C exists.* Your job is to justify each step in the proof (some of the steps require a separate RAA argument).

---

* Regarding superfluous hypotheses, there is a story that Napoleon, after examining a copy of Laplace's *Celestial Mechanics*, asked Laplace why there was no mention of God in the work. The author replied, "I have no need of this hypothesis."

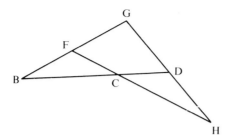

Figure 3.34

PROOF:

(1) There exists a line $\overleftrightarrow{BD}$ through B and D. *I-1*

(2) There exists a point F not lying on $\overleftrightarrow{BD}$. *I-3*

(3) There exists a line $\overleftrightarrow{BF}$ through B and F. *I-1*

(4) There exists a point G such that $B*F*G$. *B-2*

(5) Points B, F, and G are collinear. *B-1*

(6) G and D are distinct points and D, B, and G are not collinear. *RAA Argument leads to ⟹⟸ step 2*

(7) There exists a point H such that $G*D*H$.

(8) There exists a line $\overleftrightarrow{GH}$.

(9) H and F are distinct points. *RAA argument ⟹ step 2*

(10) There exists a line $\overleftrightarrow{FH}$.

(11) D does not lie on $\overleftrightarrow{FH}$. *RAA argument ∫ D∈FH ⟹ G D H F collinear But BFG collinear thus all collinear ⟹ B²*

(12) B does not lie on $\overleftrightarrow{FH}$. *RAA ⟹⟸ step 11*

(13) G does not lie on $\overleftrightarrow{FH}$.

(14) Points D, B, and G determine a triangle $\triangle DBG$ and $\overleftrightarrow{FH}$ intersects side BG in a point between B and G. *step 6 def-f △ step 4*

(15) H is the only point lying on both $\overleftrightarrow{FH}$ and $\overleftrightarrow{GH}$.

(16) No point between G and D lies on $\overleftrightarrow{FH}$. *2 lines can't ∩ in 2 pts +(G H D) doesn't hold*

(17) Hence, $\overleftrightarrow{FH}$ intersects side BD in a point C between D and B. *Pasch Thm*

(18) Thus, there exists a point C between D and B. ■

7. Given a line *l*. Fix two points A and B on it. Stipulate arbitrarily that A is *to the left* of B. Using this convention and the relation of betweenness, give a precise definition of when a point C on *l* is "to the left" of another point D on *l*. (There will be many cases to consider, differing as to how C and D are located relative to A and B.)

8. From the three-point model (p. 44) we saw that if we used only the axioms of incidence we could not prove that a line has more than two points lying on it. Using the betweenness axioms as well, prove that every line has at least five points lying on it. Give an informal argument to show that every segment (*a fortiori*, every line) has an infinite number of points lying on it (a formal proof requires the technique of mathematical induction).

9. Given a line l, a point A on l, and a point B not on l. Then every point of the ray $\overrightarrow{AB}$ (except A) is on the same side of l as B. (Hint: use an RAA argument.)

10. Prove Proposition 3.7.

11. Prove Proposition 3.8 (Hint: for Proposition 3.8c prove in two steps that E and B lie on the same side of $\overleftrightarrow{AD}$, first showing that EB does not meet $\overleftrightarrow{AD}$, then showing that EB does not meet the opposite ray $\overrightarrow{AF}$. Use Exercise 9.)

12. Prove the crossbar theorem. (Hint: assume the contrary, which means that B and C lie on the same side of $\overleftrightarrow{AD}$. Use Proposition 3.8c to derive a contradiction.)

13. Prove Proposition 3.9. (Hint: for Proposition 3.9a use Pasch's theorem and Proposition 3.7; see Figure 3.35. For Proposition 3.9b let the ray emanate from point D in the interior of $\triangle ABC$. Use the crossbar theorem and Proposition 3.7 to show that $\overrightarrow{AD}$ meets BC in a point E such that $A * D * E$. Apply Pasch's theorem to $\triangle ABE$ and $\triangle AEC$; see Figure 3.36.)

14. Prove that a line cannot be contained in the interior of a triangle.

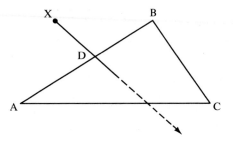

Figure 3.35

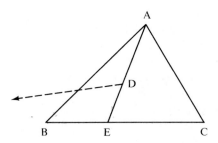

Figure 3.36

15. If a, b, and c are rays, let us say that they are *coterminal* if they emanate from the same point, and let us use the notation $a*b*c$ to mean that b is between a and c (as defined on p. 69). The analog of Axiom B-1 states that if $a*b*c$, then a, b, c are distinct, coterminal, and $c*b*a$; this analog is obviously correct. State the analogs of Axioms B-2 and B-3 and Proposition 3.3 and tell which parts of these analogs are correct. (Beware of opposite rays!)

16. Find an interpretation of the incidence axioms and the first two betweenness axioms for which Axiom B-3 fails in the following way: there exist three collinear points, no one of which is between the other two. (Hint: in the usual Euclidean model, introduce a new betweenness relation $A*B*C$ to mean that B is the midpoint of AC.)

17. Find an interpretation of the incidence axioms and the first three betweenness axioms for which the line separation property (Proposition 3.4) fails. (Hint: in the usual Euclidean model, pick a point P that is between A and B in the usual Euclidean sense and specify that A will now be considered to be between P and B. Leave all other betweenness relations among points alone. Show that P lies neither on ray $\overrightarrow{AB}$ nor its opposite ray $\overrightarrow{AC}$.)

18. A rational number of the form $a/2^n$ (with a, n integers) is called *dyadic*. In the interpretations of Major Exercise 6 below, restrict to points with dyadic coordinates and to lines through several dyadic points. Show that the incidence axioms, the first three betweenness axioms, and the line separation property all hold in the dyadic rational plane; show that Pasch's theorem fails (e.g., the lines $3x + y = 1$ and $y = 0$ do not meet in this plane).

19. A set of points S is called *convex* if whenever two points A and B are in S, the entire segment AB is contained in S. Prove that a half-plane, the interior of an angle, and the interior of a triangle are all convex sets, whereas the exterior of a triangle is not convex. Is a triangle a convex set?

Exercises on Congruence

20. Justify each step in the following proof of Proposition 3.11:

    PROOF:
    (1) Assume on the contrary that BC is not congruent to EF.
    (2) Then there is a point G on $\overrightarrow{EF}$ such that BC $\cong$ EG.
    (3) G $\neq$ F.
    (4) Since AB $\cong$ DE, adding gives AC $\cong$ DG.
    (5) However, AC $\cong$ DF.
    (6) Hence, DF $\cong$ DG.

(7) Therefore, F = G.

(8) Our assumption has led to a contradiction, hence, BC ≅ EF. ∎

21. Prove Proposition 3.13a. (Hint: in case AB and CD are not congruent, there is a unique point F ≠ D on $\overrightarrow{CD}$ such that AB ≅ CF (reason?). In case C * F * D, show that AB < CD. In case C * D * F, use Proposition 3.12 and some axioms to show that CD < AB.)

22. Use Proposition 3.12 to prove Proposition 3.13b and c.

23. Use the previous exercise and Proposition 3.3 to prove Proposition 3.13d.

24. Justify each step in the following proof of Proposition 3.14:

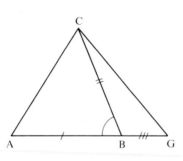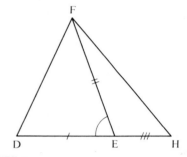

Figure 3.37

PROOF:

Given ∡ ABC ≅ ∡ DEF. To prove ∡ CBG ≅ ∡ FEH.

(1) The points A, C, and G being given arbitrarily on the sides of ∡ ABC and the supplement ∡ CBG of ∡ ABC, we can choose the points D, F, and H on the sides of the other angle and its supplement so that AB ≅ DE, CB ≅ FE, and BG ≅ EH.

(2) Then, △ABC ≅ △DEF.

(3) Hence, AC ≅ DF and ∡ A ≅ ∡ D.

(4) Also, AG ≅ DH.

(5) Hence, △ACG ≅ △DFH.

(6) Therefore, CG ≅ FH and ∡ G ≅ ∡ H.

(7) Hence, △CBG ≅ △FEH.

(8) It follows that ∡ CBG ≅ ∡ FEH, as desired. ∎

25. Deduce Proposition 3.15 from Proposition 3.14.

26. Justify each step in the following proof of Proposition 3.17 (see Figure 3.38):

PROOF:

Given △ABC and △DEF with ∡ A ≅ ∡ D, ∡ C ≅ ∡ F, and AC ≅ DF. To prove △ABC ≅ △DEF.

(1) There is a unique point B' on ray $\overrightarrow{DE}$ such that DB' $\cong$ AB.
(2) $\triangle ABC \cong \triangle DB'F$.
(3) Hence, $\sphericalangle$ DFB' $\cong \sphericalangle$ C.
(4) This implies $\overrightarrow{FE} = \overrightarrow{FB'}$.
(5) In that case, B' = E.
(6) Hence, $\triangle ABC \cong \triangle DEF$. ∎

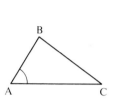
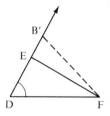

Figure 3.38

27. Prove Proposition 3.18.

28. Prove that an equiangular triangle (all angles congruent to one another) is equilateral.

29. Prove Proposition 3.20. (Hint: use Axiom C-4 and Proposition 3.19.)

30. Given $\sphericalangle$ ABC $\cong \sphericalangle$ DEF and $\overrightarrow{BG}$ between $\overrightarrow{BA}$ and $\overrightarrow{BC}$. Prove that there is a unique ray $\overrightarrow{EH}$ between $\overrightarrow{ED}$ and $\overrightarrow{EF}$ such that $\sphericalangle$ ABG $\cong \sphericalangle$ DEH. (Hint: show that D and F can be chosen so that AB $\cong$ DE and BC $\cong$ EF, and that G can be chosen so that A * G * C. Use Propositions 3.7 and 3.12 and SAS to get H. See Figure 3.25, p. 77.)

31. Prove Proposition 3.21 (imitate Exercises 21–23).

32. Prove Proposition 3.22. (Hint: use three congruence axioms to reduce to the case where A = D, C = F, and the points B and E are on opposite sides of $\overleftrightarrow{AC}$. Then consider the three cases in Figure 3.39 separately.)

33. If AB < CD, prove that 2AB < 2CD.

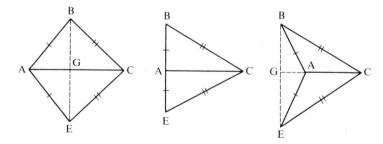

Figure 3.39

34. Let Q^2 be the *rational plane* of all ordered pairs (x, y) of rational numbers with the usual interpretations of the undefined geometric terms used in analytic geometry. Show that Axiom C-1 fails in Q^2. (Hint: the segment from $(0,0)$ to $(1, 1)$ cannot be laid off on the x axis from the origin.)

35. Let R^2 be the *real plane* of all ordered pairs (x, y) of real numbers, with the usual interpretations of the undefined geometric terms used in analytic geometry. In this plane there is a notion of the length of a segment once a standard of measurement is chosen. Let us decide to measure all lengths in inches except for segments on the x axis, which we will measure in feet, and let us *redefine congruence of segments* to mean the two segments have the same "length" in this perverse way of measuring. Incidence, betweenness, and congruence of angles will have the same meaning they usually have. Show that the first five congruence axioms still hold in this interpretation but that SAS fails (see Figure 3.40). Show that angle addition (Proposition 3.19) still holds in this interpretation.

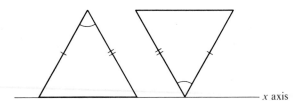

Figure 3.40

Major Exercises

1. Prove that Dedekind's axiom implies Archimedes' axiom. (Hint: divide the set of points on $\overrightarrow{AB}$ into two subsets, Σ_1 the set of all points that can be "reached" by laying off copies of segment CD starting at A, and Σ_2 the set of points that are inaccessible. Show that (Σ_1, Σ_2) is a Dedekind cut if Σ_2 is nonempty, and let O be the point of $\overleftrightarrow{AB}$ furnished by Dedekind's axiom. By an RAA argument, show that O can be reached; also by an RAA argument, show that a point to the right of O can be reached, so that Σ_2 must be empty.)

2. In order to finish the argument proving the elementary continuity principle (p. 83), i.e., to show that the point M obtained from Dedekind's axiom actually lies on the circle, we must assume that a length can be assigned to every segment and that for these lengths the *triangle inequality* holds, i.e., the sum of the lengths of two sides of a triangle exceeds the length of the third side (see Chapter 4). Using this, show that if OM < OR, there is some $M' \varepsilon \Sigma_2$ such that

OM' < OR; similarly, show that if OM > OR, there is some M'$\varepsilon\Sigma_1$ such that OM' > OR, so that we must have OM $\cong$ OR.

3. Let γ be a circle with center A and radius of length r. Let γ' be another circle with center A', radius of length r', and let d be the distance from A to A'. There is a hypothesis about the numbers r, r', and d that insures that the circles γ and γ' intersect in two distinct points. Figure out (or guess) what this hypothesis is. (Hint: it's a statement that certain numbers obtained from r, r', and d are less than certain others.)

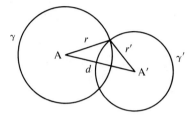

Figure 3.41

What hypothesis on r, r', and d insures that γ and γ' intersect in only one point, i.e., that the circles are tangent to each other? (See Figure 3.42.)

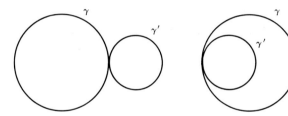

Figure 3.42

4. Report on T. L. Heath's proof for the circular continuity principle.

5. Assuming the circular continuity principle (instead of the stronger Dedekind's axiom), you can show that a line passing through the interior of a circle intersects the circle in two points; namely, in Figure 3.43, line l is assumed to pass through point A interior to circle γ with center O. Point B is taken to be the foot of the perpendicular from O to l, point C is the reflection of O across l, point D lies on γ and $\overrightarrow{OB}$. Then points E and E' are chosen collinear with O and B so that CE $\cong$ OD $\cong$ CE'. Prove that all these points satisfy the betweenness relations shown in the figure, that the circle γ' with center C and radius CE intersects γ in two points P, P', and that these points lie on the original line l. (In Exercise 17, Chapter 5, you will show that the circular continuity principle implies the elementary continuity principle).

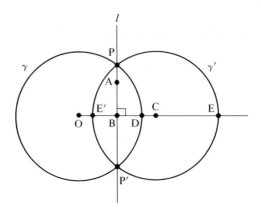

Figure 3.43

6. Incidence, points, and lines in the real plane R^2 were given in Major Exercise 9, Chapter 2. Distance is given by the usual Pythagorean formula

$$d(AB) = \sqrt{(a_1 - b_1)^2 + (a_2 - b_2)^2}$$

where $A = (a_1, a_2)$, $B = (b_1, b_2)$. Define $A * B * C$ to mean $d(AC) = d(AB) + d(BC)$, and define $AB \cong CD$ to mean $d(AB) = d(CD)$. Define $\angle ABC \cong \angle DEF$ if A, C, D, and F can be chosen on the sides of these angles so that $AB \cong ED$, $CB \cong FE$, and $AC \cong DF$. With these interpretations, verify all the axioms for Euclidean geometry (see Moise, Chapter 26, or Borsuk and Szmielew, Chapter 4).

7. Suppose in Major Exercise 6 the field R of real numbers is replaced by an arbitrary *Euclidean field* F (an ordered field in which every positive number has a square root). Show that all the axioms for Euclidean geometry except Dedekind's and Archimedes' axioms are satisfied; show also that the circular continuity principle is satisfied.

Neutral Geometry | 4

If only it could be proved ... that "there is a Triangle whose angles are together not *less* than two right angles"! But alas, *that* is an *ignis fatuus* that has never yet been caught!

C. L. Dodgson (*Lewis Carroll*)

Geometry Without the Parallel Axiom

In the exercises of the previous chapter you gained experience in proving some elementary results from Hilbert's axioms. Many of these results were taken for granted by Euclid. You can see that filling in the gaps and rigorously proving every detail is a long task. In any case, we must show that Euclid's postulates are consequences of Hilbert's. We have seen that Euclid's first postulate is the same as Hilbert's Axiom I-1. In our new language, Euclid's second postulate says the following: given segments AB and CD, there exists a point E such that A * B * E and CD ≅ BE. This follows immediately from Hilbert's Axiom C-1 applied to the ray emanating from B opposite to $\overrightarrow{BA}$.

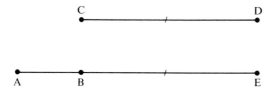

Figure 4.1

The third postulate of Euclid becomes a definition in Hilbert's system. The *circle* with *center* O and *radius* OA is defined as the set of all points P such that OP is congruent to OA. Axiom C-1 then guarantees that on every ray emanating from O there exists such a point P.

The fourth postulate of Euclid—all right angles are congruent—becomes a theorem in Hilbert's system, as was shown in Proposition 3.23.

Euclid's parallel postulate is discussed later in this chapter (p. 107 ff). In this chapter we shall be interested in neutral geometry—by definition, all those geometric theorems that can be proved using only the axioms of incidence, betweenness, congruence, and continuity and without using the axiom of parallelism. Every result proved previously is a theorem in neutral geometry. You should review all the statements in the theorems, propositions, and exercises of Chapter 3 because they will be used throughout the book. Our proofs will be less formal henceforth.

What is the purpose of studying neutral geometry? We are not interested in studying it for its own sake. Rather, we are trying to clarify the role of the parallel postulate by seeing which theorems in the geometry do not depend on it, i.e., which theorems follow from the other axioms alone without ever using the parallel postulate in proofs. This will enable us to avoid many pitfalls and to see much more clearly the logical structure of our system. Certain questions that can be answered in Euclidean geometry, e.g., whether there is a unique parallel through a given point, may not be answerable in neutral geometry because its axioms do not give us enough information.

A large number of theorems can be proved in neutral geometry (287 in Borsuk and Szmielew). We have given many familiar ones in the text and exercises of the previous chapters. Be prepared for a few strange-looking theorems in this chapter.

Alternate Interior Angle Theorem

The next theorem requires a definition: let t be transversal to lines l and l', with t meeting l at B and l' at B'. Choose points A and C on l such that $A * B * C$; choose points A' and C' on l' such that A and A' are on the same side of t and such that $A' * B' * C'$. Then the following four angles are called *interior*: $\sphericalangle A'B'B$, $\sphericalangle ABB'$, $\sphericalangle C'B'B$, $\sphericalangle CBB'$. The two pairs ($\sphericalangle ABB'$, $\sphericalangle C'B'B$) and ($\sphericalangle A'B'B$, $\sphericalangle CBB'$) are called pairs of *alternate interior angles* (see Figure 4.2).

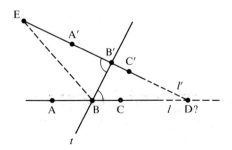

Figure 4.2

THEOREM 4.1 (Alternate Interior Angle Theorem).   If two lines cut by a transversal have a pair of congruent alternate interior angles, then the two lines are parallel.

PROOF: Given ⦠ A′B′B ≅ ⦠ CBB′. Assume on the contrary l and l' meet at a point D. Say D is on the same side of t as C and C′. There is a point E on $\overrightarrow{B'A'}$ such that B′E ≅ BD (Axiom C-1). Segment BB′ is congruent to itself, so that △B′BD ≅ △BB′E (SAS). In particular, ⦠ DB′B ≅ ⦠ EBB′. Since ⦠ DB′B is the supplement of ⦠ EB′B, ⦠ EBB′ must be the supplement of ⦠ DBB′ (Proposition 3.14 and Axiom C-4). This means that E lies on l, hence l and l' have the two points E and D in common, which contradicts Proposition 2.1 of incidence geometry. Therefore, $l \parallel l'$.■

This theorem has two very important corollaries.

COROLLARY 1.   Two lines perpendicular to the same line are parallel. Hence, the perpendicular dropped from a point P not on line l to l is *unique* (and the point at which the perpendicular intersects l is called its *foot*).

PROOF: If l and l' are both perpendicular to t, the alternate interior angles are right angles and hence are congruent (Proposition 3.23). ■

COROLLARY 2.   If l is any line and P is any point not on l, there exists at least one line m through P parallel to l.

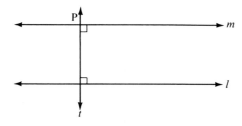

Figure 4.3

PROOF: There is a line t through P perpendicular to l, and again there is a unique line m through P perpendicular to t (Proposition 3.16). Since l and m are both perpendicular to t, Corollary 1 tells us that $l \parallel m$. (This construction will be used repeatedly.) ■

To repeat, there always exists a line *m* through P parallel to *l*—this has been proved in neutral geometry. But we don't know that *m* is *unique*. Although Hilbert's parallel postulate says that *m* is indeed unique, we are not assuming that postulate. We must keep our minds open to the strange possibility that there may be other lines through P parallel to *l*.

Warning! You are accustomed in Euclidean geometry to use the *converse* of Theorem 4.1, which states, "if two lines are parallel, then alternate interior angles cut by a transversal are congruent." We haven't proved this converse, so don't use it! (It turns out to be logically equivalent to the parallel postulate—see pp. 106–108.)

Exterior Angle Theorem

Before we continue our list of theorems, we must first make another definition: an angle supplementary to an angle of a triangle is called an *exterior angle* of the triangle; the two angles of the triangle not adjacent to this exterior angle are called the *remote interior angles*. The following theorem is a consequence of Theorem 4.1:

THEOREM 4.2 (Exterior Angle Theorem). An exterior angle of a triangle is greater than either remote interior angle [see Figure 4.4].

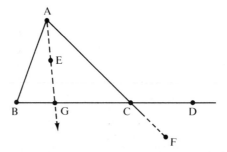

Figure 4.4

To prove $\sphericalangle$ ACD is greater than $\sphericalangle$ B and $\sphericalangle$ A:

PROOF: Consider the remote interior angle $\sphericalangle$ BAC. If $\sphericalangle$ BAC $\cong$ $\sphericalangle$ ACD, then $\overleftrightarrow{AB}$ is parallel to $\overleftrightarrow{CD}$ (Theorem 4.1), which contradicts the hypothesis that these lines meet at B. Suppose $\sphericalangle$ BAC were greater than

⊀ ACD (RAA hypothesis). Then there is a ray $\overrightarrow{AE}$ between $\overrightarrow{AB}$ and $\overrightarrow{AC}$ such that ⊀ ACD ≅ ⊀ CAE (by definition). This ray $\overrightarrow{AE}$ intersects BC in a point G (crossbar theorem, p. 69). But according to Theorem 4.1, lines $\overleftrightarrow{AE}$ and $\overleftrightarrow{CD}$ are parallel. Thus, ⊀ BAC cannot be greater than ⊀ ACD (RAA conclusion). Since ⊀ BAC is also not congruent to ⊀ ACD, ⊀ BAC must be less than ⊀ ACD (Proposition 3.21a).

For remote angle ⊀ ABC, use the same argument applied to exterior angle ⊀ BCF, which is congruent to ⊀ ACD by the vertical angle theorem (Proposition 3.15a).■

The exterior angle theorem will play a very important role in what follows. It was the sixteenth proposition in Euclid's *Elements*. Euclid's proof had a gap due to reasoning from a diagram. He considered the line $\overleftrightarrow{BM}$ joining B to the midpoint of AC and he constructed point B′ such that B * M * B′ and BM ≅ MB′ (Axiom C-1). He then assumed from the diagram that B′ lay in the interior of ⊀ ACD (see Figure 4.5).

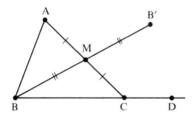

Figure 4.5

We won't give the rest of Euclid's argument (see Heath, p. 279), for there is the added difficulty that the existence of midpoints must first be justified (see Exercise 12). However, this gap in Euclid's argument can easily be filled with the tools we have developed. Namely, since segment BB′ interesects AC at M, B and B′ are on opposite sides of $\overleftrightarrow{AC}$ (by definition). Since BD meets $\overleftrightarrow{AC}$ at C, B and D are also on opposite sides of $\overleftrightarrow{AC}$. Hence, B′ and D are on the same side of $\overleftrightarrow{AC}$ (Axiom B-4). Next, B′ and M are on the same side of $\overleftrightarrow{CD}$, since segment MB′ does not contain the point B at which $\overleftrightarrow{MB'}$ meets $\overleftrightarrow{CD}$ (by construction of B′ and Axioms B-1 and B-3). Also, A and M are on the same side of $\overleftrightarrow{CD}$ because segment AM does not contain the point C at which $\overleftrightarrow{AM}$ meets $\overleftrightarrow{CD}$ (by definition of midpoint and Axiom B-3). So again, Separation Axiom B-4 insures that A and B′ are on the same side of $\overleftrightarrow{CD}$. By definition of "interior" (p. 68), we have shown that B′ lies in the interior of ⊀ ACD.

Note. In elliptic geometry the exterior angle theorem is false (see

Figure 3.24 where a triangle is shown with both an exterior angle and a remote interior angle that are right angles).

As a consequence of the exterior angle theorem (and our previous results), you can now prove as exercises the following familiar propositions (it is not necessary to use any continuity properties in your proofs).

PROPOSITION 4.1 (SAA Congruence Criterion).   Given $AC \cong DF$, $\angle A \cong \angle D$; and $\angle B \cong \angle E$. Then $\triangle ABC \cong \triangle DEF$.

Figure 4.6

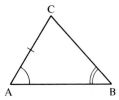
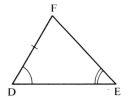

PROPOSITION 4.2.   Two right triangles are congruent if the hypotenuse and a leg of one are congruent respectively to the hypotenuse and a leg of the other.

Figure 4.7

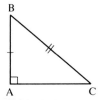
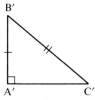

PROPOSITION 4.3 (Midpoints).   Every segment has a unique midpoint.

PROPOSITION 4.4 (Bisectors).   (a) Every angle has a unique bisector. (b) Every segment has a unique perpendicular bisector.

PROPOSITION 4.5.   In a triangle $\triangle ABC$, the greater angle lies opposite the greater side and the greater side lies opposite the greater angle, i.e., $AB > BC$ if and only if $\angle C > \angle A$.

PROPOSITION 4.6.   Given △ABC and △A'B'C', if AB ≅ A'B' and
BC ≅ B'C', then ⊰ B < ⊰ B' if and only if AC < A'C'.

Measure of Angles and Segments

Thus far in our treatment of geometry we have refrained from using
numbers that measure the sizes of angles and segments—this was in
keeping with the spirit of Euclid. From now on, however, we will not
be so austere. The next theorem (Theorem 4.3) asserts the possibility of
measurement and lists its properties. The proof requires the axioms of
continuity for the first time (in keeping with the elementary level of this
book, the interested reader is referred to Borsuk and Szmielew, Chapter 3,
§ 9 and 10).* In some popular treatments of geometry this theorem is taken
as an axiom (ruler-and-protractor postulates—see Moise). The familiar
notation (⊰ A)° will be used for the number of degrees in ⊰ A, and the
length of segment AB (with respect to some unit of measurement) will be
denoted by $\overline{AB}$.

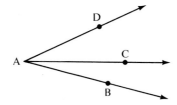

Figure 4.8

THEOREM 4.3.
A. There is a unique way of assigning a degree measure to each angle
such that the following properties hold:
(0) (⊰ A)° is a real number such that $0 \leq (\text{⊰ A})° < 180°$.
(1) (⊰ A)° = 90° if and only if ⊰ A is a right angle.
(2) (⊰ A)° = (⊰ B)° if and only if ⊰ A ≅ ⊰ B.
(3) If $\overrightarrow{AC}$ is interior to ⊰ DAB, then (⊰ DAB)° = (⊰ DAC)° + (⊰ CAB)°.

---

  * The axioms of continuity are not needed if one merely wants to define the addition for
congruence classes of segments and then prove the triangle inequality (Corollary 2 to
Theorem 4.3; see Borsuk and Szmielew, pp. 103–108, for a definition of this operation).
It is in order to prove Theorems 4.4 and 4.7 and Lemma 6.1 that we need the measurement
of angles and segments by real numbers, and for such measurement Archimedes' axiom
is required. However, parts 4 and 11 of Theorem 4.3, the proofs for which require
Dedekind's axiom, are never used in proofs in this book. See Hessenberg and Diller for
coordinatization without continuity axioms.

(4) For every real number x between 0 and 180, there exists an angle $\measuredangle$ A such that $(\measuredangle A)^\circ = x^\circ$.

(5) If $\measuredangle$ B is supplementary to $\measuredangle$ A, then $(\measuredangle A)^\circ + (\measuredangle B)^\circ = 180^\circ$.

(6) $(\measuredangle A)^\circ > (\measuredangle B)^\circ$ if and only if $\measuredangle A > \measuredangle B$.

B. Given a segment OI, called *unit segment*. Then there is a unique way of assigning a length $\overline{AB}$ to each segment AB such that the following properties hold:

(7) $\overline{AB}$ is a positive real number and $\overline{OI} = 1$.

(8) $\overline{AB} = \overline{CD}$ if and only if $AB \cong CD$.

(9) $A * B * C$ if and only if $\overline{AC} = \overline{AB} + \overline{BC}$.

(10) $\overline{AB} < \overline{CD}$ if and only if $AB < CD$.

(11) For every positive real number x, there exists a segment AB such that $\overline{AB} = x$.

Using degree notation, $\measuredangle$ A is defined as *acute* if $(\measuredangle A)^\circ < 90^\circ$, and *obtuse* if $(\measuredangle A)^\circ > 90^\circ$. Combining Theorems 4.2 and 4.3 gives the following corollary (see Exercise 2), which is essential for proving the Saccheri-Legendre theorem.

COROLLARY 1.   The sum of the degree measures of any *two* angles of a triangle is less than 180°.

The only immediate application of segment measurement that we will make is in the proof of the next corollary, the famous "triangle inequality."

COROLLARY 2 (Triangle Inequality).   If A, B, and C are three noncollinear points, then $\overline{AC} < \overline{AB} + \overline{BC}$.

PROOF:

(1) There is a unique point D such that $A * B * D$ and $BD \cong BC$ (Axiom C-1 applied to the ray opposite to $\overrightarrow{BA}$).

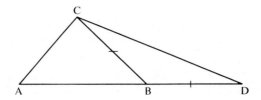

Figure 4.9

(2) Then $\measuredangle$ BCD $\cong$ $\measuredangle$ BDC (Proposition 3.10: base angles of an isosceles triangle).

(3) $\overline{AD} = \overline{AB} + \overline{BD}$ (Theorem 4.3(9)) and $\overline{BD} = \overline{BC}$ (Step 1 and Theorem 4.3(8)); substituting gives $\overline{AD} = \overline{AB} + \overline{BC}$.

(4) $\overrightarrow{CB}$ is between $\overrightarrow{CA}$ and $\overrightarrow{CD}$ (Proposition 3.7), hence, $\measuredangle$ ACD > $\measuredangle$ BCD (by definition).

(5) $\measuredangle$ ACD > $\measuredangle$ ADC (Steps 2 and 4; Proposition 3.21c).

(6) AD > AC (Proposition 4.5).

(7) Hence, $\overline{AB} + \overline{BC} > \overline{AC}$ (Theorem 4.3(10); Steps 3 and 6). ■

Saccheri-Legendre Theorem

The following very important theorem also requires an axiom of continuity (Archimedes' axiom) for its proof.

THEOREM 4.4 (Saccheri-Legendre).   The sum of the degree measures of the three angles in any triangle is *less than or equal to* 180°.

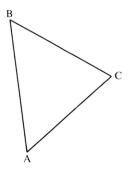

Figure 4.10

$(\measuredangle A)° + (\measuredangle B)° + (\measuredangle C)° \leqq 180°$.

This result may strike you as peculiar, since you are accustomed to the notion of an *exact* sum of 180°. Nevertheless, this exactness cannot be proved in neutral geometry! Saccheri tried, but the best he could conclude was "less than or equal." (We will discuss this further in this chapter, pp. 108–112.) Max Dehn showed in 1900 that there is no way to prove this theorem without Archimedes' axiom.* The idea of the proof is as follows:

Assume, on the contrary, that the angle sum of $\triangle ABC$ is greater than 180°, say $180° + p°$, where p is a positive number. It is possible (by a trick you will find in Exercise 15) to replace $\triangle ABC$ with another triangle that has the same angle sum as $\triangle ABC$ but in which one angle has at most half the number of degrees as $(\nleqslant A)°$. We can repeat this trick to get another triangle that has the same angle sum $180° + p°$ but in which one angle has at most one-fourth the number of degrees as $(\nleqslant A)°$. The Archimedean property of real numbers guarantees that if we repeat this construction enough times, we will eventually obtain a triangle that has angle sum $180° + p°$ but in which one angle has degree measure at most $p°$. The sum of the degree measures of the other *two* angles will be greater than or equal to 180°, contradicting Corollary 1 to Theorem 4.3. This proves the theorem.

You should prove the following consequence of the Saccheri-Legendre theorem as an exercise.

COROLLARY 1.   The sum of the degree measures of two angles in a triangle is less than or equal to the degree measure of their remote exterior angle [see Figure 4.11].

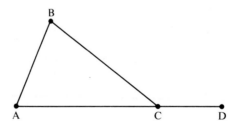

Figure 4.11
$(\nleqslant A)° + (\nleqslant B)° \leqq (\nleqslant BCD)°$.

It is natural to generalize the Saccheri-Legendre theorem to polygons other than triangles. For example, let us prove that the angle sum of a quadrilateral $\square ABCD$ is at most 360°. Break $\square ABCD$ into two triangles

---

* See the reference in Major Exercise 1. The full significance of Archimedes' axiom was first grasped in the 1880s by M. Pasch and O. Stolz. G. Veronese and T. Levi-Civita developed the first non-Archimedean geometry.

△ABC and △ADC by the diagonal AC (see Figure 4.12). By the Saccheri-Legendre theorem,

$$(\measuredangle B)° + (\measuredangle BAC)° + (\measuredangle ACB)° \leqq 180° \quad \text{and}$$
$$(\measuredangle D)° + (\measuredangle DAC)° + (\measuredangle ACD)° \leqq 180°.$$

Theorem 4.3(3) gives us the equations

$$(\measuredangle BAC)° + (\measuredangle DAC)° = (\measuredangle BAD)° \quad \text{and}$$
$$(\measuredangle ACB)° + (\measuredangle ACD)° = (\measuredangle BCD)°$$

Using these equations, we add the two inequalities to obtain the desired inequality

$$(\measuredangle B)° + (\measuredangle D)° + (\measuredangle BAD)° + (\measuredangle BCD)° \leqq 360°.$$

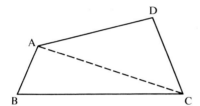

Figure 4.12

Unfortunately, there is a gap in this simple argument! To get the equations used above, we assumed by looking at the diagram (Figure 4.12) that C was interior to $\measuredangle$ BAD and that A was interior to $\measuredangle$ BCD. But what if the quadrilateral looked like Figure 4.13?

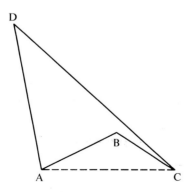

Figure 4.13

In this case the equations would not hold. To prevent such a case, we must add an hypothesis; we must assume that the quadrilateral is "convex."

DEFINITION.   Quadrilateral □ABCD is called *convex* if it has a pair of opposite sides, e.g., AB and CD, such that CD is contained in one of the half-planes bounded by $\overleftrightarrow{AB}$ and AB is contained in one of the half-planes bounded by $\overleftrightarrow{CD}$.*

The assumption made above is now justified by starting with a convex quadrilateral. Thus, we have proved the following corollary:

COROLLARY 2.   The sum of the degree measures of the angles in any *convex* quadrilateral is at most 360°.

Note. The Saccheri-Legendre theorem is false in elliptic geometry (see Figure 3.24, p. 76). In fact, it can be proved in elliptic geometry that the angle sum of a triangle is always greater than 180°.

Equivalence of Parallel Postulates

We shall now prove the equivalence of Euclid's fifth postulate and Hilbert's parallel postulate. Note, however, that we are not proving either or both of the postulates; we are only proving that we *can* prove one *if* we first assume the other. We shall first state Euclid V (all the terms in the statement have now been defined carefully).

---

* It can be proved that this condition also holds for the other pair of opposite sides, AD and BC—see Exercise 23 in this chapter. The use of the word "convex" in this definition does not agree with its use in Exercise 19, Chapter 3; a convex quadrilateral is obviously not a "convex set" as defined in that exercise. However, we can define the *interior* of a convex quadrilateral □ABCD as follows: each side of □ABCD determines a half-plane containing the opposite side; the interior of □ABCD is then the intersection of the four half-planes so determined. You can then prove that the interior of a convex quadrilateral is a convex set (which is one of the problems in Exercise 25).

EUCLID'S POSTULATE V.   If two lines are intersected by a transversal in such a way that the sum of the degree measures of the two interior angles on one side of the transversal is less than 180°, then the two lines meet on that side of the transversal.

THEOREM 4.5.   Euclid's fifth postulate ⇔ Hilbert's parallel postulate.

PROOF :   First, assume Hilbert's postulate. The situation of Euclid V is shown in Figure 4.14. $(\not\!\prec 1)° + (\not\!\prec 2)° < 180°$ (hypothesis) and $(\not\!\prec 1)° + (\not\!\prec 3)° = 180°$ (supplementary angles, Theorem 4.3(5)). Hence, $(\not\!\prec 2)° < 180° - (\not\!\prec 1)° = (\not\!\prec 3)°$. There is a unique ray $\overrightarrow{B'C}$ such that $\not\!\prec 3$ and $\not\!\prec$ C'B'B are congruent alternate interior angles (Axiom C-4). By Theorem 4.1, $\overleftrightarrow{B'C}$ is parallel to l. Since $m \neq \overleftrightarrow{B'C}$, m meets l (Hilbert's postulate). To conclude, we must prove that m meets l on the same side of t as C'. Assume, on the contrary, that they meet at a point A on the opposite side. Then $\not\!\prec 2$ is an exterior angle of $\triangle$ABB'. Yet it is smaller than the remote interior $\not\!\prec$ 3. This contradiction of Theorem 4.2 proves Euclid V (RAA).

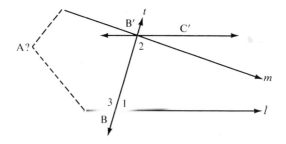

Figure 4.14

Conversely, assume Euclid V and refer to Figure 4.15, the situation of Hilbert's postulate. Let t be the perpendicular to l through P, and m the perpendicular to t through P. We know that $m \| l$ (Corollary 1 to Theorem 4.1). Let n be any other line through P. We must show that n meets l. Let $\not\!\prec$ 1 be the acute angle n makes with t (which angle exists because $n \neq m$). Then $(\not\!\prec 1)° + (\not\!\prec \text{PQR})° < 90° + 90° = 180°$. Thus, the hypothesis of Euclid V is satisfied. Hence, n meets l, proving Hilbert's postulate. ■

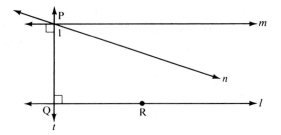

Figure 4.15

Since Hilbert's parallel postulate and Euclid V are logically equivalent in the context of neutral geometry, Theorem 4.5 allows us to use them interchangeably. You will prove as exercises that the following statements are also logically equivalent to the parallel postulate.

PROPOSITION 4.7.   Hilbert's parallel postulate ⇔ if a line intersects one of two parallel lines, then it also intersects the other.

PROPOSITION 4.8.   Hilbert's parallel postulate ⇔ converse to Theorem 4.1 (alternate interior angles).

PROPOSITION 4.9.   Hilbert's parallel postulate ⇔ if t is transversal to l and m, $l \parallel m$, and $t \perp l$, then $t \perp m$.

PROPOSITION 4.10.   Hilbert's parallel postulate ⇔ if $k \parallel l$, $m \perp k$, and $n \perp l$, then either $m = n$ or $m \parallel n$.

The next proposition is another statement logically equivalent to Hilbert's parallel postulate, but at this point we can only prove the implication in one direction (the other implication is proved in Chapter 6).

PROPOSITION 4.11.   Hilbert's parallel postulate ⇒ the angle sum of every triangle is 180°.

Angle Sum of a Triangle

We define the *angle sum* of triangle $\triangle ABC$ as $(\angle A)° + (\angle B)° + (\angle C)°$, which is a certain number of degrees $\leq 180°$ (by the Saccheri-Legendre theorem). We define the *defect* of any triangle to be 180° minus the angle sum. In Euclidean geometry we are accustomed to having no "defective" triangles, i.e. we are accustomed to having the defect equal zero (Proposition 4.11).

The main purpose of this section is to show that if *one* defective triangle exists, then *all* triangles are defective. Or, put in the contrapositive form, if one triangle has angle sum 180°, then so do all others. We are not asserting that one such triangle does exist, nor are we asserting the contrary; we are only examining the hypothesis that one might exist.

THEOREM 4.6.   Let △ABC be any triangle and D a point between A and B. Then defect (△ABC) = defect (△ACD) + defect (△BCD) (*additivity of the defect*).

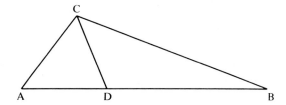

Figure 4.16

PROOF: Since $\overrightarrow{CD}$ is interior to ⧡ACB (Proposition 3.7), $(⧡ACB)° = (⧡ACD)° + (⧡BCD)°$ (by Theorem 4.3(3)). Since ⧡ADC and ⧡BDC are supplementary angles, $180° = (⧡ADC)° + (⧡BDC)°$ (by Theorem 4.3(5)). To obtain the additivity of the defect, all we have to do is write down the definition of the defect (180° minus the angle sum) for each of the three triangles under consideration and substitute the two equations above (Exercise 1). ∎

COROLLARY.   Under the same hypothesis, the angle sum of △ABC is equal to 180° if and only if the angle sums each of △ACD and △BCD are equal to 180°.

PROOF: If △ACD and △BCD both have defect zero, then defect of △ABC = 0 + 0 = 0 (Theorem 4.6). Conversely, if △ABC has defect zero, then by Theorem 4.6 defect (△ACD) + defect (△BCD) = 0. But the defect of a triangle can never be negative (Saccheri-Legendre theorem). Hence, △ACD and △BCD each have defect zero (the sum of two non-negative numbers equals zero only when each equals zero). ∎

Next, recall that by definition a *rectangle* is a quadrilateral whose four angles are right angles. Hence, the angle sum of a rectangle is 360°. Of course, we don't yet know whether rectangles exist in neutral geometry. (Try to construct one without using the parallel postulate or any statement logically equivalent to it—see Exercise 19.)

The next theorem is the result we seek. Its proof will be given in five steps.

THEOREM 4.7.   If a triangle exists whose angle sum is 180°, then a rectangle exists. If a rectangle exists, then every triangle has angle sum equal to 180°.

PROOF :
(1) Construct a *right* triangle having angle sum 180°.

Let △ABC be the given triangle with defect zero (hypothesis). Assume it is not a right triangle, otherwise we are done. At least two of the angles in this triangle are acute, since the angle sum of two angles in a triangle must be less than 180° (Corollary to Theorem 4.3), e.g., ∡ A and ∡ B are acute. Let CD be the altitude from vertex C (which exists by Proposition 3.16). We claim that D lies between A and B. Assume the contrary, e.g., D ∗ A ∗ B.

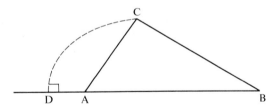

Figure 4.17

Then remote interior angle ∡ CDA is greater than exterior angle ∡ CAB, contradicting Theorem 4.2. Similarly, if A ∗ B ∗ D, we get a contradiction. Thus, A ∗ D ∗ B (Axiom B-3).

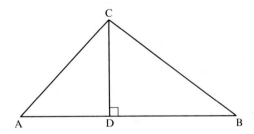

Figure 4.18

It now follows from the Corollary to Theorem 4.6 that each of the right triangles △ADC and △BDC have defect zero.

(2) From a right triangle of defect zero construct a rectangle.

Let △CDB be a right triangle of defect zero with ∡ D a right angle. By Axiom C-4, there is a unique ray $\overrightarrow{CX}$ on the opposite side of $\overleftrightarrow{CB}$ from D such that ∡ DBC ≅ ∡ BCX. By Axiom C-1, there is a unique point E on $\overrightarrow{CX}$ such that CE ≅ BD.

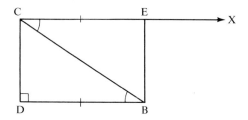

Figure 4.19

Then △CDB ≅ △BEC (SAS). Hence, △BEC is also a right triangle of defect zero with right angle at E. Also, since (∡ DBC)° + (∡ BCD)° = 90° by our hypothesis, we obtain by substitution (∡ ECB)° + (∡ BCD)° = 90°   and   (∡ DBC)° + (∡ EBC)° = 90°. Moreover, B is an interior point of ∡ ECD, since the alternate interior angle theorem implies $\overleftrightarrow{CE} \parallel \overleftrightarrow{DB}$ and $\overleftrightarrow{CD} \parallel \overleftrightarrow{BE}$ and C is interior to ∡ EBD (for the same reason). Thus, we can apply Theorem 4.3(3) to conclude that (∡ ECD)° = 90° = (∡ EBD)°. This proves that □CDBE is a rectangle.

(3) From one rectangle, construct "arbitrarily large" rectangles.

More precisely, given any right triangle △D'E'C', construct a rectangle □AFBC such that AC > D'C' and BC > E'C'. This can be done using Archimedes' axiom. We simply "lay off" enough copies of the rectangle we have to achieve the result (see Figures 4.20 and 4.21; you can make this "laying off" precise as an exercise).

(4) Prove that all *right* triangles have defect zero.

This is achieved by "embedding" an arbitrary right triangle △D'C'E' in a rectangle, as in Step 3, then showing successively (by twice applying the Corollary to Theorem 4.6) that △ACB, △DCB, and △DCE each have defect zero (see Figure 4.22).

(5) If every *right* triangle has defect zero, then *every* triangle has defect zero.

As in Step 1, drop an altitude to decompose an arbitrary triangle into two right triangles (Figure 4.18) and apply the corollary to Theorem 4.6.■

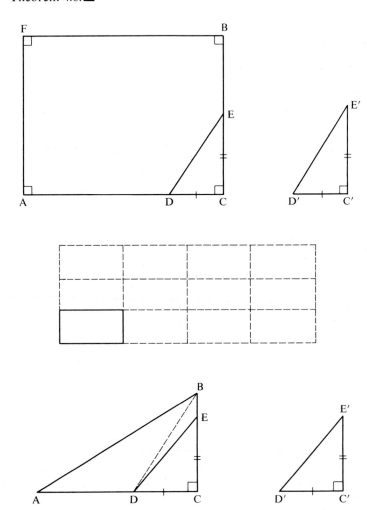

Figure 4.20

Figure 4.21

Figure 4.22

Historians credit Theorem 4.7 to Saccheri and Legendre, but we will not name it after them so as to avoid confusion with Theorem 4.4.

COROLLARY. If there exists a triangle with positive defect, then all triangles have positive defect.

Review Exercise

Which of the following statements are correct?

(1) If two triangles have the same defect, they are congruent.

(2) Euclid's fourth postulate is a theorem in neutral geometry.

(3) Theorem 4.5 shows that Euclid's fifth postulate is a theorem in neutral geometry.

(4) The Saccheri-Legendre theorem tells us that some triangles exist that have angle sum less than 180° and some triangles exist that have angle sum equal to 180°.

(5) The alternate interior angle theorem states that if parallel lines are cut by a transversal, then alternate interior angles are congruent to each other.

(6) It is impossible to prove in neutral geometry that quadrilaterals exist.

(7) The Saccheri-Legendre theorem is false in Euclidean geometry because in Euclidean geometry the angle sum of any triangle is never less than 180°.

(8) According to our definition of "angle," the degree measure of an angle cannot equal 180°.

(9) The notion of one ray being "between" two others is undefined.

(10) It is impossible to prove in neutral geometry that parallel lines exist.

(11) The definition of "remote interior angle" given on p. 98 is incomplete because it used the word "adjacent," which has never been defined.

(12) An exterior angle of a triangle is any angle that is not in the interior of the triangle.

(13) The SSS criterion for congruence of triangles is a theorem in neutral geometry.

(14) The alternate interior angle theorem implies, as a special case, that if a transversal is perpendicular to one of two parallel lines, then it is also perpendicular to the other.

(15) Another way of stating the Saccheri-Legendre theorem is to say that the defect of a triangle cannot be negative.

(16) The ASA criterion for congruence of triangles is one of the axioms for neutral geometry.

(17) The proof of Theorem 4.7 depends on Archimedes' axiom.

(18) If $\triangle ABC$ is any triangle and C is any of its vertices, and if a perpendicular is dropped from C to $\overleftrightarrow{AB}$, then that perpendicular will intersect $\overleftrightarrow{AB}$ in a point between A and B.

(19) It is a theorem in neutral geometry that given any point P and any line l, there is at most one line through P perpendicular to l.

(20) It is a theorem in neutral geometry that vertical angles are congruent to each other.

(21) The proof of Theorem 4.2 (on exterior angles) uses Theorem 4.1 (on alternate interior angles).

(22) The gap in Euclid's attempt to prove Theorem 4.2 can be filled using our axioms of betweenness.

Exercises

The following are exercises in neutral geometry, unless otherwise stated. This means that in your proofs you are allowed to use only those results that have been given previously (including results from previous exercises). You are not allowed to use the parallel postulate or other results from Euclidean geometry that depend on it.

1. (a) Finish the last step in the proof of Theorem 4.6. (b) Prove that congruent triangles have the same defect. (c) Prove the corollary to Theorem 4.7 (p. 110).

2. Prove Corollary 1 to Theorem 4.3 (p. 102) by applying Theorem 4.2 and parts 5 and 6 of Theorem 4.3.

3. State the converse to Euclid's fifth postulate. Prove this converse as a theorem in neutral geometry.

4. Prove Propostion 4.7.

5. Prove Proposition 4.8. (Hint: assume the converse to Theorem 4.1. Let m be the parallel to l through P constructed in the proof of Corollary 2 to Theorem 4.1 and let n be any parallel to l through P. Use the congruence of alternate interior angles and the uniqueness of perpendiculars to prove $m = n$. Assuming next the parallel postulate, use Axiom C-4 and an RAA argument to establish the converse to Theorem 4.1.)

6. Prove Proposition 4.9.

7. Prove Proposition 4.10.

8. Prove Proposition 4.11. (Hint: see Figure 4.23.)

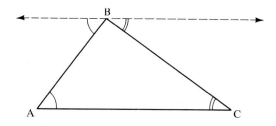

Figure 4.23

9. The following purports to be a proof in neutral geometry of the SAA criterion for congruence. Find the flaw.

   Given AC ≅ DE, ∡ A ≅ ∡ D, ∡ B ≅ ∡ E. Then ∡ C ≅ ∡ F, since $(\angle C)° = 180° − (\angle A)° − (\angle B)° = 180° − (\angle D)° − (\angle E)° = (\angle F)°$ (Theorem 4.3(2)). Hence, △ABC ≅ △DEF by ASA (Proposition 3.17).

10. Here is a correct proof of the SAA criterion. Justify each step. (1) Assume side AB is not congruent to side DE. (2) Then AB < DE or DE < AB. (3) If DE < AB, then there is a point G between A and B such that AG ≅ DE (see Figure 4.24). (4) Then △CAG ≅ △FDE. (5) Hence, ∡ AGC ≅ ∡ E. (6) It follows that ∡ AGC ≅ ∡ B. (7) This contradicts a certain theorem (which?). (8) Therefore, DE is not less than AB. (9) By a similar argument involving a point H between D and E, AB is not less than DE. (10) Hence, AB ≅ DE. (11) Therefore, △ABC ≅ △DEF.

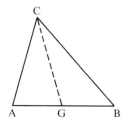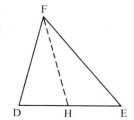

Figure 4.24

11. Prove Proposition 4.2. (Hint: see Figure 4.7, p. 100. On the ray opposite to $\overrightarrow{AC}$, lay off segment AD congruent to A′C′. First prove △DAB ≅ △C′A′B′, then use isosceles triangles and the SAA criterion to conclude.)

12. Here is a proof that segment AB has a midpoint. Justify each step (see Figure 4.25).

    (1) Let C be any point not on $\overleftrightarrow{AB}$. (2) There is a unique ray $\overrightarrow{BX}$ on the opposite side of $\overleftrightarrow{AB}$ from C such that ∡ CAB ≅ ∡ ABX. (3) There is a unique point D on $\overrightarrow{BX}$ such that AC ≅ BD. (4) D is on the opposite side of $\overleftrightarrow{AB}$ from C. (5) Let E be the point at which segment CD intersects $\overleftrightarrow{AB}$. (6) Assume E is

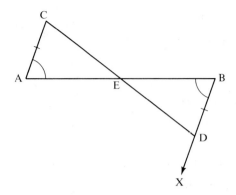

Figure 4.25

not between A and B. (7) Then either E = A, or E = B, or E ∗ A ∗ B, or A ∗ B ∗ E. (8) $\overleftrightarrow{AC}$ is parallel to $\overleftrightarrow{BD}$. (9) Hence, E ≠ A and E ≠ B. (10) Assume E ∗ A ∗ B (Figure 4.26).

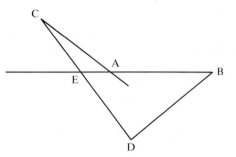

Figure 4.26

(11) Since $\overleftrightarrow{CA}$ intersects side EB of △EBD at a point between E and B, it must also intersect either ED or BD. (12) Yet this is impossible. (13) Hence, A is not between E and B. (14) Similarly, B is not between A and E. (15) Thus, A ∗ E ∗ B (see Figure 4.25). (16) Then ⊀ AEC ≅ ⊀ BED. (17) △EAC ≅ △EBD. (18) Therefore, E is a midpoint of AB.

13. Prove that segment AB has only one midpoint. (Hint: assume the contrary, and use Propositions 3.3, 3.5, and 3.13 to derive a contradiction.)

14. Prove Proposition 4.4 (on bisectors). (Hint: use midpoints.)

15. Prove the following result, needed to demonstrate the Saccheri-Legendre theorem (see Figure 4.27). Let D be the midpoint of BC and E the unique point on $\overleftrightarrow{AD}$ such that A ∗ D ∗ E and AD ≅ DE. Then △AEC has the same angle sum as △ABC, and either (⊀ EAC)° or (⊀ AEC)° is $\leq \frac{1}{2}$ (⊀ BAC)°. (Hint: first show that △BDA ≅ △CDE, then that (⊀ EAC)° + (⊀ AEC)° = (⊀ BAC)°.)

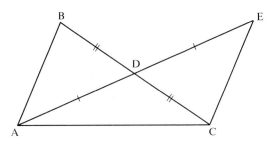

Figure 4.27

16. Prove Corollary 1 to the Saccheri-Legendre theorem.

17. Prove the following theorems:
    (a) Let γ be a circle with center O, and let A and B be two points on γ. The segment AB is called a *chord* of γ; let M be its midpoint. Then $\overleftrightarrow{OM}$ is perpendicular to $\overleftrightarrow{AB}$. (Hint: corresponding angles of congruent triangles are congruent.)
    (b) Let AB be a chord of the circle γ having center O. Prove that the perpendicular bisector of AB passes through the center O of γ.

18. A familiar Euclidean statement is "an angle inscribed in a semicircle is a right angle." Prove that this statement implies the existence of a right triangle $\triangle ABC$ whose angle sum is 180°.

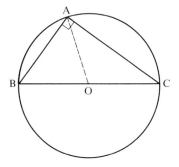

Figure 4.28

19. Find the flaw in the following argument purporting to construct a rectangle. Let A and B be any two points. There is a unique line l through A perpendicular to $\overleftrightarrow{AB}$ (Proposition 3.16) and, similarly, there is a unique line m through B perpendicular to $\overleftrightarrow{AB}$. Take any point C on m other than B. There is a unique line through C perpendicular to l—let it intersect l at D. Then $\square ABCD$ is a rectangle.

20. On p. 99 we saw how the gap in Euclid's proof of Theorem 4.2 (on exterior angles) could be filled, and in Exercise 12 we justified Euclid's use of a midpoint. Now finish Euclid's argument to obtain another proof of Theorem 4.2.

21. Prove Proposition 4.5. (Hint: if $AB > BC$, then let D be the point between A and B such that $BD \cong BC$ (Figure 4.29). Use isosceles triangle $\triangle CBD$ and exterior angle $\angle BDC$ to show that $\angle ACB > \angle A$. Use this result and trichotomy of ordering to prove the converse.)

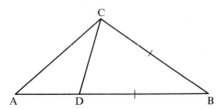

Figure 4.29

22. Prove Proposition 4.6. (Hint: given $\angle B < \angle B'$. Use the hypothesis of Proposition 4.6 to reduce to the case $A = A'$, $B = B'$, and C interior to $\angle ABC'$, so that you must show $AC < AC'$ (see Figure 4.30). This is easy in case $C = D$, where point D is obtained from the crossbar theorem. In case $C \neq D$, Proposition 4.5 reduces the problem to showing that $\angle AC'C < \angle ACC'$. In

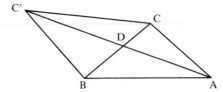

Figure 4.30
$BC \cong BC'$.

case $B * D * C$ (as in Figure 4.30), you can prove this inequality using the congruence $\angle BCC' \cong \angle BC'C$. In case $B * C * D$ (Figure 4.31), apply the congruence $\angle BCC' \cong \angle BC'C$ and Theorem 4.2 to exterior angle $\angle BCC'$ of $\triangle DCC'$ and exterior angle $\angle DCC'$ of $\triangle BCC'$. (The converse implication in Proposition 4.6 follows from the direct implication, just shown, if you apply trichotomy.)

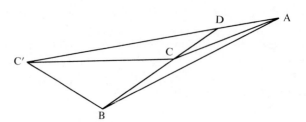

Figure 4.31

23. For the purpose of this exercise, call segments AB and CD *semiparallel* if segment AB does not intersect line $\overleftrightarrow{CD}$ and segment CD does not intersect line $\overleftrightarrow{AB}$. Obviously, if $\overleftrightarrow{AB} \| \overleftrightarrow{CD}$, then AB and CD are semiparallel, but the converse need not hold (see Figure 4.32). We have defined a quadrilateral to be convex if one pair of opposite sides is semiparallel. Prove that the other pair of opposite sides is also semiparallel. (Hint: suppose AB is semiparallel to CD and assume, on the contrary, that AD meets $\overleftrightarrow{BC}$ in a point E. Use the definition of quadrilateral (Exercise 3, Chapter 1) to show either that $E * B * C$ or $B * C * E$; in either case, use Pasch's theorem, p. 67, to derive a contradiction.)

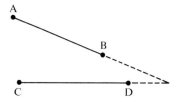

Figure 4.32

24. Prove that the diagonals of a convex quadrilateral intersect. (Hint: apply the crossbar theorem.)

25. Prove that the intersection of convex sets (defined in Exercise 19, Chapter 3) is again a convex set. Use this result to prove that the interior of a convex quadrilateral is a convex set and that the point at which the diagonals intersect lies in the interior.

26. The *convex hull* of a set of points S is the intersection of all the convex sets containing S, i.e., it is the smallest convex set containing S. Prove that the convex hull of three noncollinear points A, B, and C consists of the sides and interior of $\triangle ABC$.

27. Given $A * B * C$ and $\overleftrightarrow{DC} \perp \overleftrightarrow{AC}$. Prove that $AD > BD > CD$ (Figure 4.33; use Proposition 4.5).

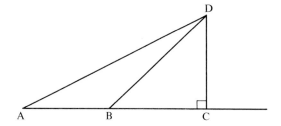

Figure 4.33

28. Given any triangle $\triangle DAC$ and any point B between A and C. Prove that either $DB < DA$ or $DB < DC$. (Hint: drop a perpendicular from D to $\overleftrightarrow{AC}$ and use the previous exercise.)

29. Prove that the interior of a circle is a convex set. (Hint: use the previous exercise.)

30. Prove that if D is an exterior point of △ABC, then there is a line $\overleftrightarrow{DE}$ through D that is contained in the exterior of △ABC (see Figure 4.34).

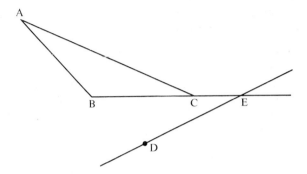

Figure 4.34

31. Suppose that line *l* meets circle γ in two points C and D. Prove that:
    (a) Point P on *l* lies inside γ if and only if C * P * D.
    (b) If points A and B are inside γ and on opposite sides of *l*, then the point E at which AB meets *l* is between C and D.

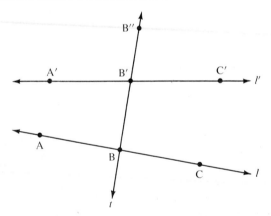

Figure 4.35

32. In Figure 4.35, the pairs of angles (⋌ A'B'B″, ⋌ ABB″) and (⋌ C'B'B″, ⋌ CBB″) are called pairs of *corresponding angles* cut off on *l* and *l'* by transversal *t*. Prove that corresponding angles are congruent if and only if alternate interior angles are congruent.

Major Exercises

1. Read Chapter 28 of E. E. Moise's *Elementary Geometry from an Advanced Standpoint*, which describes an ordered field that is not Archimedean. Show that

if degrees are measured by numbers in such a field, then the proof of the Saccheri-Legendre theorem (sketched on p. 104) breaks down. If you can read German, look up the paper by M. Dehn, "Die Legendreschen Sätze über die Winkelsumme im Dreieck," *Mathematische Annalen*, Bd. 53 (1900), pp. 405–439, and report on it. (Dehn exhibits two non-Archimedean models of the incidence, betweenness and congruence axioms, one in which the Saccheri-Legendre theorem fails and another in which the converse to Proposition 4.11 (p. 108) fails.)

2. Consider the following statements on congruence:
   1. Given triangle △ABC and segment DE such that AB ≅ DE. Then on a given side of $\overleftrightarrow{DE}$ there is a unique point F such that AC ≅ DF and BC ≅ EF.
   2. Given triangles △ADC and △A′D′C′ and given A∗B∗C and A′∗B′∗C′. If AB ≅ A′B′, BC ≅ B′C′, AD ≅ A′D′, and BD ≅ B′D′, then CD ≅ C′D′ ("rigidity of a triangle with a tail"—see Figure 4.36).

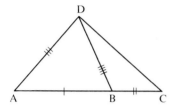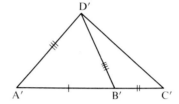

Figure 4.36

Prove these statements. Also, prove a statement 2a obtained from statement 2 by substituting CD ≅ C′D′ for BD ≅ B′D′ in the hypothesis and making BD ≅ B′D′ the conclusion.

In Borsuk and Szmielew, statements 1 and 2 are taken as axioms, in place of our Axioms C-4, C-5, and C-6. The advantage of this change is that these new congruence axioms refer only to congruence of segments. Congruence of angles, ∢ABC ≅ ∢A′B′C′, can then be *defined* by specifying that A and C (resp. A′ and C′) can be chosen on the sides of ∢B (resp. ∢B′) so that AB ≅ A′B′, BC ≅ B′C′, and AC ≅ A′C′. With this definition, keeping the same incidence and betweenness axioms as before, show that C-4, C-5, and C-6 can be proved from C-1, C-2, C-3, and statements 1 and 2. (Hint: first prove statement 2a by an RAA argument. Then show that if ∢ABC ≅ ∢A′B′C′, and that if we had chosen other points D, E, D′, and E′ on the sides of ∢B and ∢B′ such that DB ≅ D′B′ and EB ≅ E′B′, then DE ≅ D′E′. See Figure 4.37.)

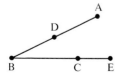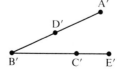

Figure 4.37

History of the Parallel Postulate

5

Like the goblin "Puck," [the feat of proving Euclid] has led me "up and down, up and down," through many a wakeful night: but always, just as I thought I had it, some unforeseen fallacy was sure to trip me up, and the tricksy sprite would "leap out, laughing ho, ho, ho!"

C. L. Dodgson (Lewis Carroll)

Let us summarize what we have done so far. We have discovered certain gaps in Euclid's definitions and postulates for plane geometry. We filled in these gaps and firmed up the foundations for this geometry by presenting (a modified version of) Hilbert's definitions and axioms. We then built a structure of theorems on these foundations. However, the structure thus far erected does not rest on the parallel postulate, and we called that structure "neutral geometry." One reason we postponed building on the parallel postulate is that we have less confidence in it than in the other axioms.

Here is Euclid's phrasing of his fifth postulate: "That, if a straight line falling on two straight lines make the interior angles on the same side less than two right angles, the two straight lines, if produced indefinitely, meet on that side on which are the angles less than two right angles."

You may think that this postulate is more complicated than the others, and yet you may also feel that to deny it would go against common sense. Albert Einstein once said that "common sense is, as a matter of fact, nothing more than layers of preconceived notions stored in our memories and emotions for the most part before age eighteen."

That Euclid himself did not quite trust this postulate is shown by the fact that he postponed using it in a proof for as long as possible—until his twenty-ninth proposition. The postulate was also doubted by the Greeks of Euclid's time and in the centuries following him.

Proclus

Proclus (410–485 A.D.), whose commentary is one of the main sources of information on Greek geometry, criticized the parallel postulate as follows: "This ought even to be struck out of the Postulates altogether; for it is a theorem involving many difficulties, which Ptolemy, in a certain book, set himself to solve. . . . The statement that since [the two lines] converge more and more as they are produced, they will sometime meet is plausible but not necessary." Proclus offers the example of a hyperbola that approaches its asymptotes as closely as you like without ever meeting them. This example shows that the opposite of Euclid's conclusions can at least be imagined. Proclus says: "It is then clear from this that we must seek a proof of the present theorem, and that it is alien to the special character of postulates."

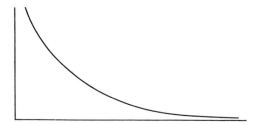

Figure 5.1

For over two thousand years some of the best mathematicians tried to prove Euclid's fifth postulate. What does it mean, according to our terminology, to have a proof? It should not be necessary to *assume* the parallel postulate as an axiom; we should be able to prove it *from the other axioms.* If we were able to prove Euclid V in this way, it would become a theorem in neutral geometry and neutral geometry would encompass all of Euclidean geometry.

The first known attempted proof was by Ptolemy. Without going through the details of his argument (see Heath, pp. 204–206), we might say that he assumed Hilbert's parallel postulate without realizing it. We have seen in Chapter 4 (pp. 106–108) that Hilbert's parallel postulate is logically equivalent to Euclid V (Theorem 4.5), so that Ptolemy was assuming what he was trying to prove, i.e., his reasoning was essentially circular.

Proclus attempted to prove the parallel postulate as follows: Given two parallel lines l and m. Suppose line n cuts m at P. We wish to show n intersects l also (cf. Proposition 4.7). Let Q be the foot of the perpendicular from P to l (Corollary 1 to Theorem 4.1). If n coincides with $\overleftrightarrow{PQ}$, then it intersects l at Q. Otherwise, one ray $\overrightarrow{PY}$ of n lies between $\overrightarrow{PQ}$ and a ray $\overrightarrow{PX}$ of m (see Exercise 15, Chapter 3). Take X to be the foot of the perpendicular from Y to m.

Now, as the point Y recedes endlessly from P on n, the segment XY increases indefinitely in size, and so eventually becomes greater than

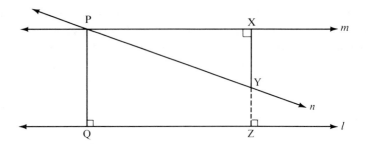

Figure 5.2

segment PQ. Therefore, Y must cross over to the other side of *l*, so that *n* must meet *l*.

The preceding paragraph is the heart of Proclus' argument; it is a rather sophisticated argument, involving motion and continuity. Moreover, every step in the argument can be shown to be correct—except that the conclusion doesn't follow! (In Exercise 6 you are asked to prove that XY increases indefinitely, where "indefinitely" means "without bound." For example, the sequence of numbers $\frac{1}{2}, \frac{3}{4}, \frac{7}{8}, \frac{15}{16}, \frac{31}{32}, \ldots$ increases but not "indefinitely" in the sense of "without bound," because the numbers are always less than one; one is a bound for these numbers.)

How could one justify the last step? Let us drop a perpendicular YZ from Y to *l*. You might then say that (1) X, Y, and Z are collinear, and (2) XZ $\cong$ PQ. Thus, when XY becomes greater than PQ, XY must also be greater than XZ, so that Y must be on the other side of *l*. Here the conclusion does indeed follow from statements (1) and (2). The trouble is that there is no justification for these statements!

If this boggles your mind, it may be because Figure 5.2 makes statements (1) and (2) *seem* correct. You recall, however, that we are not allowed to use a diagram to justify a step in a proof. Each step must be proved from stated axioms or previously proven theorems. (We will show later that it is not possible in neutral geometry to prove statements (1) and (2). They can be proved only in Euclidean geometry and only by using the parallel postulate; this makes Proclus' argument circular.)

This analysis of Proclus' faulty argument illustrates how careful you must be in the way you think about parallel lines from now on. You probably visualize parallel lines as railroad tracks, everywhere equidistant from each other, and the ties of the tracks perpendicular to both parallels. This imagery is valid only in Euclidean geometry. Without the parallel postulate, the only thing we can say about two lines that are "parallel" is that, by definition of "parallel," they have no point in common. You can't assume they are equidistant; you can't even assume they have one common perpendicular segment. As Humpty Dumpty remarked: "When I use a word it means what I wish it to mean, neither more nor less."

Wallis

The next important attempt to prove the parallel postulate was made by the Persian astronomer and mathematician Nasiraddin (1201–1274). But since his attempted proof had several unjustified assumptions, let us move ahead to John Wallis (1616–1703).* Wallis gave up trying to prove the parallel postulate in neutral geometry. Instead, he proposed a new axiom, which he felt was more plausible than the parallel postulate, and then proved the parallel postulate from his new axiom and the other axioms of neutral geometry.

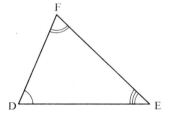

Figure 5.3

WALLIS' POSTULATE : Given any triangle △ABC and given any segment DE. There exists a triangle △DEF (having DE as one of its sides) that is similar to △ABC (denoted △DEF ~ △ABC).

Recall that *similar triangles* are triangles whose vertices can be put in one-to-one correspondence so that corresponding angles are congruent. In

---

* Wallis was the leading English mathematician before Isaac Newton. In his treatise *Arithmetica infinitorum* (which Newton studied), Wallis introduced the symbol ∞ for "infinity," developed formulas for certain integrals, and presented his famous infinite product formula

$$\frac{\pi}{2} = \frac{2 \cdot 2 \cdot 4 \cdot 4 \cdot 6 \cdot 6 \cdot 8 \cdots}{1 \cdot 3 \cdot 3 \cdot 5 \cdot 5 \cdot 7 \cdot 7 \cdots}.$$

Euclidean geometry it is proved that corresponding sides of similar triangles are proportional (see Exercise 18), e.g., each side of △DEF might be twice as long as the corresponding side of △ABC. Thus, the intuitive meaning of Wallis' postulate is that you can either magnify or shrink a triangle as much as you like, without distortion.

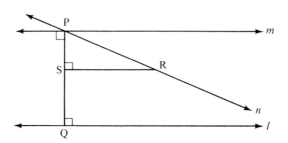

Figure 5.4

Using Wallis' postulate, the parallel postulate can be proved as follows:

PROOF: Given a point P not on line *l*, construct one parallel *m* to *l* through P as before—by dropping a perpendicular $\overleftrightarrow{PQ}$ to *l* and erecting *m* perpendicular to $\overleftrightarrow{PQ}$. Let *n* be any other line through P. We must show that *n* meets *l*. As before, we consider a ray of *n* emanating from P that is between a ray of *m* and $\overrightarrow{PQ}$; for any point R on this ray, we drop $\overleftrightarrow{RS}$ perpendicular to $\overleftrightarrow{PQ}$ (cf. Proposition 3.16 and Corollary 1 to Theorem 4.1 for the existence and uniqueness of all our perpendiculars).

We now apply Wallis' postulate to △PSR and segment PQ. It tells us that there is a point T such that △PSR is similar to △PQT. Moreover, let T lie on the same side of $\overleftrightarrow{PQ}$ as R (Exercise 7).

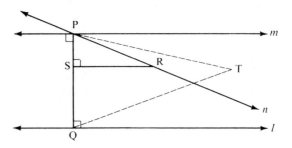

Figure 5.5

By definition of similar triangles, $\not\subset$ TPQ $\cong$ $\not\subset$ RPS. But since these angles have the ray $\overrightarrow{PQ} = \overrightarrow{PS}$ as a common side and since T lies on the same side of $\overleftrightarrow{PQ}$ as R, the only way they can be congruent is to be equal (Axiom C-4). Thus, $\overrightarrow{PR} = \overrightarrow{PT}$, so that T lies on n. Similarly, $\not\subset$ PQT $\cong$ $\not\subset$ PSR, a right angle; hence, T lies on l. Thus, n and l meet at T; m is therefore the only line through P parallel to l.■

There is no longer reason to consider Wallis' postulate any more plausible than Euclid V, because it turns out to be logically equivalent to Euclid V (see Exercise 8).

Saccheri and Lambert

In Exercises 9–11 we will discuss other attempts to prove the parallel postulate, notably the incorrect proofs by Legendre and Wolfgang Bolyai. Here we should mention the work of the Jesuit priest Girolamo Saccheri (1667–1733). Just before he died he published a little book entitled *Euclides ab omni naevo vindicatus* (Euclid Freed of Every Flaw), which was not really noticed until a century and a half later, when Eugenio Beltrami rediscovered it.

Saccheri's idea was to use a reductio ad absurdum argument. He assumed the negation of the parallel postulate and tried to deduce a contradiction. Specifically, he studied certain quadrilaterals (Figure 5.6) whose base angles are right angles and whose sides are congruent to each other. These quadrilaterals have subsequently become known as *Saccheri quadrilaterals*. It is easy to prove in neutral geometry that the summit angles are congruent (Exercise 1), i.e., $\not\subset$ C $\cong$ $\not\subset$ D.

There are three possible cases:

Case 1: The summit angles are right angles.
Case 2: The summit angles are obtuse.
Case 3: The summit angles are acute.

Figure 5.6

Johann Heinrich Lambert

Wanting to prove the first case, which is the case in Euclidean geometry, Saccheri tried to show that the other two cases led to contradictions. He succeeded in showing that Case 2 leads to a contradiction: if the summit angles were obtuse, the angle sum of the quadrilateral would be more than 360°, contradicting Corollary 2 to the Saccheri-Legendre theorem (p. 106; to verify the hypothesis of Corollary 2, see Exercise 14).

However hard he tried, he could not squeeze a contradiction out of Case 3, "the inimical acute angle hypothesis," as he called it. He was able to deduce many strange results, but not a contradiction. Finally, he exclaimed in frustration: "The hypothesis of the acute angle is absolutely false, because [it is] repugnant to the nature of the straight line!" It is as if a man had discovered a rare diamond, but, unable to believe what he saw, announced it was glass. Although he did not recognize it, Saccheri had discovered non-Euclidean geometry.

In a similar approach to the parallel problem, Johann Heinrich Lambert (1728–1777) studied quadrilaterals having at least three right angles, which are now named after him (see Exercise 4). He, too, deduced many non-Euclidean propositions from the acute angle hypothesis but, unlike Saccheri, he did not claim to have found a contradiction. Going beyond Saccheri, Lambert showed that the acute angle hypothesis implies that the area of a triangle is proportional to its defect, and he speculated that this hypothesis corresponded to the geometry on a "sphere of imaginary radius."*

Wolfgang Bolyai

There were so many attempts to prove Euclid V that by 1763 G. S. Klügel was able to submit a doctoral thesis finding the flaws in 28 different supposed proofs of the parallel postulate, expressing doubt that it could be proved. The French encyclopedist and mathematician J. L. R. d'Alembert called this "the scandal of geometry." Even as late as 1823 the great French mathematician Legendre thought he had a proof (we presented one of his attempts in Chapter 1, pp. 19–21). Mathematicians

---

* Beltrami later showed that the geometry on a *pseudosphere* satisfies the acute angle hypothesis (see Appendix A). Lambert is also known for developing hyperbolic trigonometry and for proving that when x is a rational number, e^x and $\tan x$ are irrational.

Wolfgang (Farkas) Bolyai

were becoming discouraged. The Hungarian Wolfgang (Farkas) Bolyai wrote to his son János:

> You must not attempt this approach to parallels. I know this way to its very end. I have traversed this bottomless night, which extinguished all light and joy of my life. I entreat you, leave the science of parallels alone. . . . I thought I would sacrifice myself for the sake of the truth. I was ready to become a martyr who would remove the flaw from geometry and return it purified to mankind. I accomplished monstrous, enormous labors; my creations are far better than those of others and yet I have not achieved complete satisfaction. . . . I turned back when I saw that no man can reach the bottom of the night. I turned back unconsoled, pitying myself and all mankind.
>
> I admit that I expect little from the deviation of your lines. It seems to me that I have been in these regions; that I have traveled past all reefs of this infernal Dead Sea and have always come back with broken mast and torn sail. The ruin of my disposition and my fall date back to this time. I thoughtlessly risked my life and happiness—*aut Caesar aut nihil*.*

But the young Bolyai was not deterred by his father's warnings, for he had a completely new idea. He assumed that the negation of Euclid's parallel postulate was not absurd, and in 1823 was able to write to his father:

> It is now my definite plan to publish a work on parallels as soon as I can complete and arrange the material and an opportunity presents itself; at the moment I still do not clearly see my way through, but the path which I have followed gives positive evidence that the goal will be reached, if it is at all possible; I have not quite reached it, but I have discovered such wonderful things that I was amazed, and it would be an everlasting piece of bad fortune if they were lost. When you, my dear Father, see them, you will understand; at present I can say nothing except this: that *out of nothing I have created a strange new universe*. All that I have sent you previously is like a house of cards in comparison with a tower. I am no less convinced that these discoveries will bring me honor than I would be if they were completed.*

We will explore this "strange new universe" in the following chapters. A century after János Bolyai wrote this letter, the English physicist J. J. Thomson remarked, somewhat facetiously:

> We have Einstein's space, de Sitter's space, expanding universes, contracting universes, vibrating universes, mysterious universes. In fact, the pure mathe-

* Meschkowski, 1964.

matician may create universes just by writing down an equation, and indeed if he is an individualist he can have a universe of his own.

In fact, in 1949 the renowned logician Kurt Gödel found a model of the universe that satisfies Einstein's gravitational equations, one in which it is theoretically possible to travel backwards in time!*

Review Exercise

Which of the following statements are correct?

(1) Wallis' postulate implies that there exist two triangles that are similar but not congruent.

(2) A "Saccheri quadrilateral" is a quadrilateral □ABDC such that ∡ CAB and ∡ DBA are right angles and AC ≅ BD.

(3) A "Lambert quadrilateral" is a quadrilateral having at least three right angles.

(4) A quadrilateral that is both Saccheri and Lambert must be a rectangle.

(5) A hyperbola comes arbitrarily close to its asymptotes without ever intersecting them.

(6) János Bolyai warned his son Wolfgang not to work on the parallel problem.

(7) Saccheri succeeded in disproving the "inimical" acute angle hypothesis.

(8) In trying to prove Euclid's fifth postulate, Ptolemy tacitly assumed what we have been calling Hilbert's parallel postulate.

(9) It is a theorem in neutral geometry that if $l \| m$ and $m \| n$, then $l \| n$.

(10) It is a theorem in neutral geometry that every segment has a unique midpoint.

(11) It is a theorem in neutral geometry that if a rectangle exists, then the angle sum of any triangle is 180°.

(12) It is a theorem in neutral geometry that if l and m are parallel lines, then alternate interior angles cut out by any transversal to l and m are congruent to each other.

---

* To date, attempts to refute Gödel's model on either mathematical or philosophical grounds have failed. See "On the paradoxical time-structures of Gödel," by Howard Stein, *Journal of the Philosophy of Science,* December 1970, and a forthcoming paper by Paul Horwich.

Exercises

Again, in proofs in Exercises 1–17 you are allowed to use only our previous results from neutral geometry.

1. Let □ABDC be a Saccheri quadrilateral, so that ∢ B and ∢ A are right angles and CA ≅ DB. Prove that ∢ C ≅ ∢ D. (Hint: prove △CAB ≅ △DBA, then △CDB ≅ △DCA.)

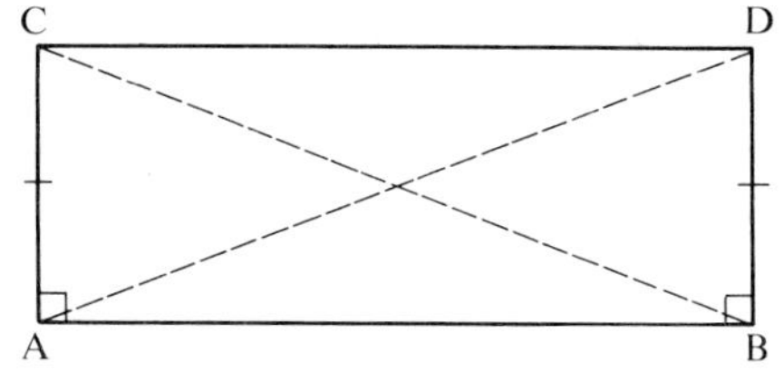

Figure 5.7

2. Let □ABDC be a quadrilateral whose base angles ∢ A and ∢ B are right angles. Prove that if AC < BD, then (∢ D)° < (∢ C)°. (Hint: if AC ≅ BE, with B ∗ E ∗ D, use Exercise 23, Chapter 4, to show that E is interior to ∢ ACD, then apply Exercise 1 and the exterior angle theorem, etc.)

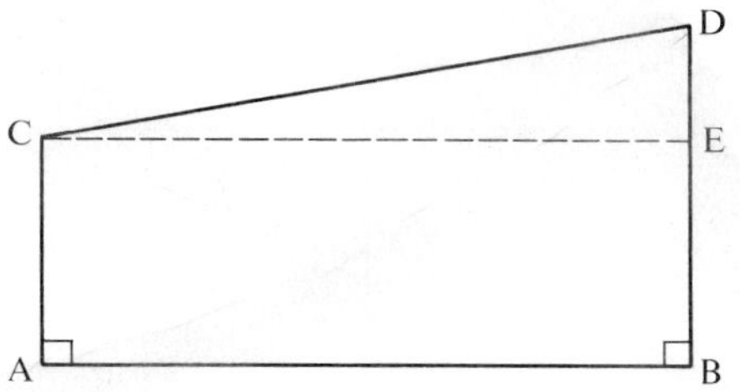

Figure 5.8

3. With the same hypothesis as in Exercise 2, prove the converse, that if (∢ D)° < (∢ C)°, then AC < BD. (Hint: assume the contrary, which involves the two cases AC ≅ BD and AC > BD. In each case, derive a contradiction.)

4. The Swiss-German mathematician Lambert considered quadrilaterals with at least three right angles which are now named after him.

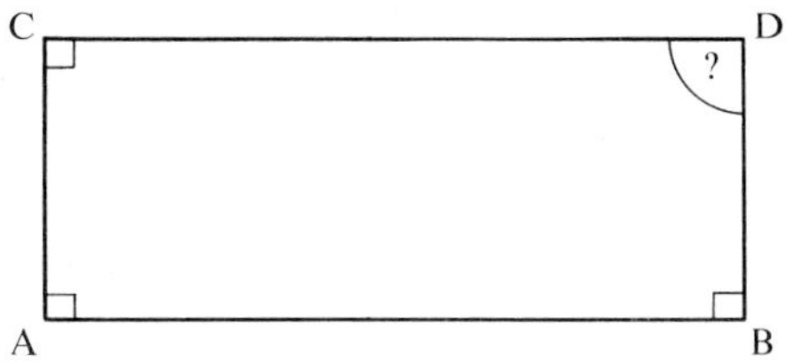

Figure 5.9

Prove the following:

(a) The fourth angle ∢ D of a Lambert quadrilateral is never obtuse.
(b) If ∢ D is a right angle, then the opposite sides of □ABCD are congruent (use Exercise 2 and an RAA argument).
(c) If ∢ D is acute, then a side adjacent to ∢ D is greater than its opposite side, i.e., DB > CA and CD > AB (use Exercise 3).
(d) A quadrilateral is both Lambert and Saccheri if and only if it is a rectangle.

We can combine statements (a), (b), and (c) of this exercise into the following statement: A side adjacent to the fourth angle of a Lambert quadrilateral is greater than or congruent to its opposite side. As you know, case (b) always holds if the geometry is Euclidean; in the next chapter we will show that case (c) always holds if the geometry is hyperbolic. In elliptic geometry the fourth angle of a Lambert quadrilateral is always obtuse, and a side adjacent to the fourth angle is always smaller than its opposite side.

5. Given a right triangle △PXY with right angle at X, form a new right triangle △PX′Y′ that has acute angle ∢ P in common with the given triangle but double the hypotenuse (prove that this can be done). Prove that the side opposite the acute angle is *at least* doubled, whereas the side adjacent to the acute angle is *at most* doubled. (Hint: extend side XY far enough to drop a perpendicular Y′Z to $\overleftrightarrow{XY}$. Prove that △PXY ≅ △Y′ZY, and apply Exercise 4 to the Lambert quadrilateral □XZY′X′.)

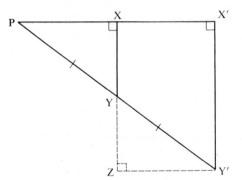

Figure 5.10

6. Use Exercise 5 to justify the next to last statement in Proclus' argument, that as Y recedes endlessly from P, segment XY increases indefinitely (cf. p. 123). (Hint: use Archimedes' axiom and the fact that $2^n \to \infty$ as $n \to \infty$.) Does segment PX also increase indefinitely?

7. Given a right triangle △TQP with right angle at Q. Prove that there is a unique point T′ on $\overleftrightarrow{TQ}$ such that T′ is on the opposite side of $\overleftrightarrow{PQ}$ from T and △T′QP ≅ △TQP (see Figure 5.11).

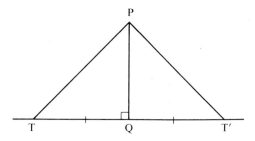

Figure 5.11

8. Prove that Euclid's fifth postulate implies Wallis' postulate. (Hint: use also Axiom C-4 and the fact that in Euclidean geometry the angle sum of a triangle is 180°—Proposition 4.11.)

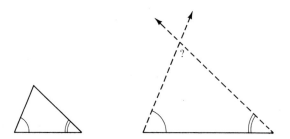

Figure 5.12

9. Find the unjustified assumption in the following "proof" of the parallel postulate by Wolfgang Bolyai: Given P not on line *l*, $\overleftrightarrow{PQ}$ perpendicular to *l* at Q, and line *m* perpendicular to $\overleftrightarrow{PQ}$ at P. Let *n* be any line through P distinct from *m* and $\overleftrightarrow{PQ}$. We must show that *n* meets *l*. Let A be any point between P and

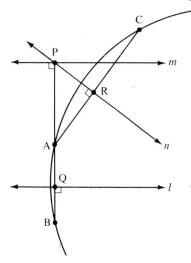

Figure 5.13

Q. Let B be the unique point such that A ∗ Q ∗ B and AQ ≅ QB. Let R be the foot of the perpendicular from A to *n*. Let C be the unique point such that A ∗ R ∗ C and AR ≅ RC. Then, A, B, and C are not collinear, hence there is a unique circle γ passing through them. Since *l* is the perpendicular bisector of chord AB of γ, and *n* is the perpendicular bisector of chord AC of γ, then *l* and *n* meet at the center of γ (Exercise 17b, Chapter 4).

10. Reread Legendre's attempted proof of the parallel postulate in Chapter 1 (pp. 19–21). Find the flaw and justify all the steps that are correct.

11. The following attempted proof of the parallel postulate is similar to Proclus' but the flaw is different; detect the flaw with the help of Exercise 5. Given P not on line *l*, $\overleftrightarrow{PQ}$ perpendicular to *l* at Q, and line *m* perpendicular to $\overleftrightarrow{PQ}$ at P. Let *n* be any line through P distinct from *m* and $\overleftrightarrow{PQ}$. We must show that *n* meets *l*. Let $\overrightarrow{PX}$ be a ray of *n* between $\overrightarrow{PQ}$ and a ray of *m*, and let Y be the foot of the perpendicular from X to $\overleftrightarrow{PQ}$. As X recedes endlessly from P, PY increases indefinitely. Hence, Y eventually reaches a position Y′ on $\overrightarrow{PQ}$ such that PY′ > PQ. Let X′ be the corresponding position reached by X on line *n*. Now X′ and Y′ are on the same side of *l* because $\overleftrightarrow{X'Y'}$ is parallel to *l*. But Y′ and P are on opposite sides of *l*. Hence, X′ and P are on opposite sides of *l*, so that segment PX′ (which is part of *n*) meets *l*.

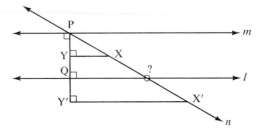

Figure 5.14

12. Find the flaw in the following attempted proof of the parallel postulate given by J. D. Gergonne: Given P not on line *l*, $\overleftrightarrow{PQ}$ perpendicular to *l* at Q, line *m* perpendicular to $\overleftrightarrow{PQ}$ at P, and point A ≠ P on *m*. Let $\overrightarrow{PB}$ be the last ray between $\overrightarrow{PA}$ and $\overrightarrow{PQ}$ that intersects *l*, B being the point of intersection. There exists a point C on *l* such that Q ∗ B ∗ C (Axioms B-1 and B-2). It follows that

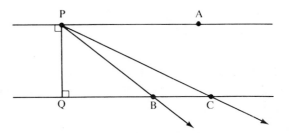

Figure 5.15

$\vec{PB}$ is not the last ray between $\vec{PA}$ and $\vec{PQ}$ that intersects l, hence all rays between $\vec{PA}$ and $\vec{PQ}$ meet l. Thus m is the only parallel to l through P.

13. Suppose that in the statement of Wallis' postulate we add the assumption AB $\cong$ DE and replace the word "similar" by "congruent." Prove this new statement to be a theorem in neutral geometry.

14. In our diagram of a Saccheri quadrilateral $\square$ABDC having base AB we drew points C and D on the same side of $\overleftrightarrow{AB}$, but this condition is not part of the definition. Justify this diagram by proving that a Saccheri quadrilateral is a convex quadrilateral (see p. 106 for the definition).

15. A quadrilateral is called a *parallelogram* if the lines containing opposite sides are parallel. Prove that a Lambert quadrilateral is a parallelogram and that every parallelogram is a convex quadrilateral.

16. Prove that if line l cuts circle γ at A and is not tangent to γ at A then l meets γ at a second point. (Hint: reflect A across the perpendicular t from the center of γ to l.)

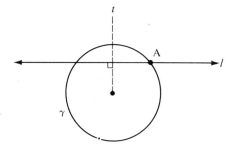

Figure 5.16

17. Prove that the circular continuity principle implies the elementary continuity principle (p. 82). (Hint: apply Proposition 3.4, Major Exercise 5, Chapter 3, and Exercises 27 and 31a, Chapter 4.)

The remaining exercises in this chapter are exercises in Euclidean geometry, which means you are allowed to use the parallel postulate and its consequences already established. We will refer to these results in Chapter 7. You are also allowed to use the following result, a proof of which is indicated in the major exercises:

PARALLEL PROJECTION THEOREM.   Given three parallel lines l, m, and n. Let t and t' be transversals to these parallels, cutting them in points A, B, and C and in points A', B', and C', respectively. Then $\overline{AB}/\overline{BC} = \overline{A'B'}/\overline{B'C'}$.

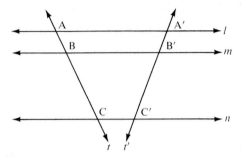

Figure 5.17

18. *Fundamental Theorem on Similar Triangles.* Given △ABC ~ △A′B′C′, i.e., given ∡ A ≅ ∡ A′, ∡ B ≅ ∡ B′ and ∡ C ≅ ∡ C′. Then corresponding sides are proportional, i.e., $\overline{AB}/\overline{A'B'} = \overline{AC}/\overline{A'C'} = \overline{BC}/\overline{B'C'}$. Prove the theorem. (Hint: let B″ be the point on $\overrightarrow{AB}$ such that AB″ ≅ A′B′, and let C″ be the point on $\overrightarrow{AC}$ such that AC″ ≅ A′C′. Use the hypothesis to show that △AB″C″ ≅ △A′B′C′ and deduce from corresponding angles that $\overleftrightarrow{B''C''}$ is parallel to $\overleftrightarrow{BC}$. Now apply the parallel projection theorem.)

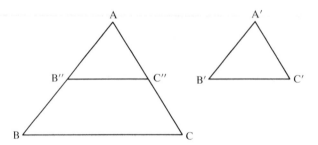

Figure 5.18

19. Prove the converse to the fundamental theorem on similar triangles. (Hint: choose B″ as before. Use Pasch's theorem to show that the parallel to $\overleftrightarrow{BC}$ through B″ cuts AC at a point C″. Then use the hypothesis, Exercise 18, and the SSS criterion to show that △ABC ~ △AB″C″ ≅ △A′B′C′.)

20. *SAS Similarity Criterion.* If ∡ A ≅ ∡ A′ and $\overline{AB}/\overline{A'B'} = \overline{AC}/\overline{A'C'}$, prove that △ABC ~ △A′B′C′. (Hint: same method as in Exercise 19, but using SAS instead of SSS.)

21. Prove the Pythagorean theorem. (Hint: let CD be the altitude to the hypotenuse. Use the fact that the angle sum of a triangle equals 180° (Exercise 8, Chapter 4) to show that △ACD ~ △ABC ~ △CBD. Apply Exercise 18 and a little algebra based on $\overline{AB} = \overline{AD} + \overline{DB}$ to get the result.)

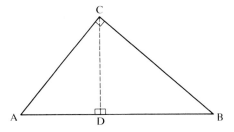

Figure 5.19

22. The fundamental theorem on similar triangles (Exercise 18) allows the trigonometric functions sine, cosine, etc., to be defined. Namely, given an acute angle
∢ A, make it part of a right triangle △BAC with right angle at C, and set

$$\sin \measuredangle A = (\overline{BC})/(\overline{AB})$$
$$\cos \measuredangle A = (\overline{AC})/(\overline{AB}).$$

These definitions are then independent of the choice of the right triangle used.
If ∢ A is obtuse and ∢ A′ is its supplement, set

$$\sin \measuredangle A = + \sin \measuredangle A'$$
$$\cos \measuredangle A = - \cos \measuredangle A'.$$

If ∢ A is a right angle, set

$$\sin \measuredangle A = 1$$
$$\cos \measuredangle A = 0.$$

Now, given any triangle △ABC, if a and b are the lengths of the sides opposite
A and B, respectively, prove the law of sines,

$$\frac{a}{b} = \frac{\sin \measuredangle A}{\sin \measuredangle B}.$$

(Hint: drop altitude CD and use the two right triangles △ADC and △BDC
to show that $b \sin \measuredangle A = \overline{CD} = a \sin \measuredangle B$.) Similarly, prove the law of cosines

$$c^2 = a^2 + b^2 - 2ab \cos \measuredangle C$$

and deduce the converse to the Pythagorean theorem.

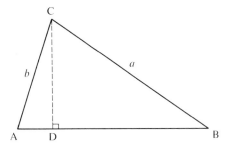

Figure 5.20

23. Given $A * B * C$ and point D not collinear with A, B, and C. Prove that

$$\frac{\overline{AB}}{\overline{BC}} = \frac{\overline{AD} \sin \sphericalangle ADB}{\overline{CD} \sin \sphericalangle CDB}$$

$$\frac{\overline{AC}}{\overline{BC}} = \frac{\overline{AD} \sin \sphericalangle ADC}{\overline{BD} \sin \sphericalangle BDC}.$$

(Hint: use the law of sines to compute $\overline{AB}/\overline{AD}$, $\overline{CD}/\overline{BC}$, and $\overline{BD}/\overline{BC}$, and remember that $\sin \sphericalangle ABD = \sin \sphericalangle CBD$.)

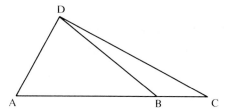

Figure 5.21

24. Let γ be a circle with center O, and let P, Q, and R be three points on γ. Prove that if P and R are diametrically opposite, then $\sphericalangle$ PQR is a right angle, and otherwise $(\sphericalangle PQR)° = \frac{1}{2} (\sphericalangle POR)°$. (Hint: again use the fact that triangular angle sum is 180°. There are four cases to consider, as in Figure 5.22.)

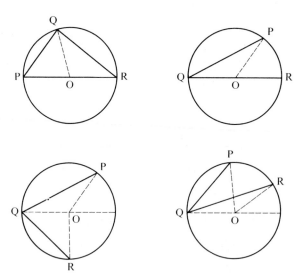

Figure 5.32

25. Prove that if two angles inscribed in a circle subtend the same arc, then they are congruent. (Hint: apply the previous exercise.)

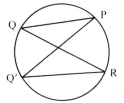

Figure 5.23

∢ PQR ≅ ∢ PQ′R.

26. Prove that if ∢ PQR is a right angle, then Q lies on the circle γ having PR as diameter. (Hint: use uniqueness of perpendiculars and Exercises 16 and 24.)

Major Exercises

These exercises furnish the proof of the parallel projection theorem in Euclidean geometry (p. 137; also see Figure 5.17).

1. Prove that in Euclidean geometry opposite sides of a parallelogram are congruent to each other.

2. Let k, l, m, and n be parallel lines, distinct, except that possibly $l = m$. Let transversals t and t' cut these lines in points A, B, C, and D and in A′, B′, C′, and D′, respectively. If AB ≅ CD, prove that A′B′ ≅ C′D′. (Hint: construct parallels to t through A′ and C′. Apply Major Exercise 1 and the congruence of corresponding angles.)

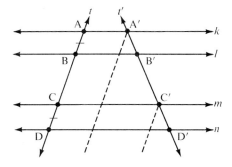

Figure 5.24

3. Prove that parallel projection preserves betweenness, i.e., in Figure 5.17, if $A * B * C$, then $A' * B' * C'$. (Hint: use Axiom B-4.)

4. Prove the parallel projection theorem for the special case in which the ratio of lengths $\overline{AB}/\overline{BC}$ is a rational number p/q. (Hint: divide AB into p congruent segments and BC into q congruent segments so that all $p + q$ segments will be congruent. Apply Major Exercise 2 $p + q$ times.)

5. The case where $\overline{AB}/\overline{BC}$ is an irrational number x is the difficult case. Let $\overline{A'B'}/\overline{B'C'} = x'$. The idea is to show that every rational number p/q less than x is also less than x' (and by symmetry, vice versa). This will imply $x = x'$, since a real number is the least upper bound of all the rational numbers less than it. To show this, lay off on $\overrightarrow{BC}$ a segment BD of length $p\overline{AB}/q$, and let D′ be the parallel projection of D onto t'. From $p/q < x$, deduce $B * D * C$. Now apply Major Exercises 3 and 4 to show that $p/q < x'$.*

6. Given a segment AB of length a with respect to some unit segment OI (see Theorem 4.3). Using straightedge and compass only, show how to construct a segment of length $\sqrt{a}$. (Hint: extend AB to a segment AC of length $a + 1$, erect a perpendicular through B and let D be one of its intersections with the circle having AC as diameter; apply the theory of similar triangles to show that $\overline{BD} = \sqrt{a}$. Review the construction in Major Exercise 1, Chapter 1.)

---

* This clever method of proof was essentially discovered by the ancient Greek mathematician Eudoxus—see E. C. Zeeman, "Research, Ancient and Modern," to appear in the *Bulletin of the Institute of Mathematics and Its Applications,* Warwick University, England.

The Discovery of Non-Euclidean Geometry

6

> Out of nothing I have created a strange new universe.
>
> *János Bolyai*

János Bolyai

It is remarkable that sometimes when the time is right for a new idea to come forth, the idea occurs to several people more or less simultaneously. Thus it was in the eighteenth century with the discovery of the calculus by Newton in England and Leibniz in Germany, and in the nineteenth century with the discovery of non-Euclidean geometry. When János Bolyai (1802–1860) announced privately his discoveries in non-Euclidean geometry, his father Wolfgang admonished him:

> It seems to me advisable, if you have actually succeeded in obtaining a solution of the problem, that, for a two-fold reason, its publication be hastened: first, because ideas easily pass from one to another who, in that case, can publish them; secondly, because it seems to be true that many things have, as it were, an epoch in which they are discovered in several places simultaneously, just as the violets appear on all sides in springtime.*

János Bolyai did publish his discoveries, as a 26-page appendix to a book by his father (the *Tentamen*, 1831). His father eagerly sent a copy of this book to his friend, the German mathematician Carl Friedrich Gauss (1777–1855), undisputedly the foremost mathematician of his time. Wolfgang Bolyai had become close friends with Gauss 35 years earlier, when they were both students in Göttingen. After Wolfgang returned to Hungary, they maintained an intimate correspondence† and when Wolfgang sent Gauss his own attempt to prove the parallel postulate, Gauss tactfully pointed out the fatal flaw.

---

* Quoted in Meschkowski, 1964. For evidence that non-Euclidean ideas were "in the air" during the early nineteenth century, see the account of the contributions of F. L. Wachter, C. F. Schweikart, and Taurinus in H. E. Wolfe, *Non-Euclidean Geometry*, pp. 56–60 (Holt, Rinehart, Winston, 1945).

† See Schmidt and Stäckel, 1972, for the complete correspondence (in German).

János was 13 years old when he mastered the differential and integral calculus. His father wrote to Gauss begging him to take the young prodigy into his household as an apprentice mathematician. Gauss never replied to this request (perhaps because he was having enough trouble with his own son Eugene, who had run away from home). Fifteen years later, when Wolfgang mailed the *Tentamen* to Gauss, he certainly must have felt that his son had vindicated his belief in him, and János must have expected Gauss to publicize his achievement. One can therefore imagine the disappointment János must have felt when he read the following letter to his father from Gauss:

If I begin with the statement that I dare not praise such a work, you will of course be startled for a moment: but I cannot do otherwise; to praise it would amount to praising myself; for the entire content of the work, the path which your son has taken, the results to which he is led, coincide almost exactly with my own meditations which have occupied my mind for from thirty to thirty-five years. On this account I find myself surprised to the extreme.

My intention was, in regard to my own work, of which very little up to the present has been published, not to allow it to become known during my lifetime. Most people have not the insight to understand our conclusions and I have encountered only a few who received with any particular interest what I communicated to them. In order to understand these things, one must first have a keen perception of what is needed, and upon this point the majority are quite confused. On the other hand, it was my plan to put all down on paper eventually, so that at least it would not finally perish with me.

So I am greatly surprised to be spared this effort, and am overjoyed that it happens to be the son of my old friend who outstrips me in such a remarkable way.

Despite the compliment in Gauss' last sentence, János was bitterly disappointed with the great mathematician's reply; he even imagined that his father had secretly informed Gauss of his results and that Gauss was now trying to appropriate them as his own. A man of fiery temperament, who had fought and won thirteen successive duels (unlike Galois, who was killed in a duel at age 20), János fell into deep mental depression and never again published his research. A translation of his immortal "appendix" can be found in R. Bonola's *Non-Euclidean Geometry* (1955).

Gauss

There is evidence that Gauss had anticipated some of J. Bolyai's discoveries, in fact, that Gauss had been working on non-Euclidean geometry since the age of 15, i.e., since 1792 (see Bonola, Chapter 3). In 1817, Gauss wrote to W. Olbers: "I am becoming more and more convinced that the necessity of our [Euclidean] geometry cannot be proved, at least not by human reason nor for human reason. Perhaps in another life we will be able to obtain insight into the nature of space, which is now inattainable." In 1824, Gauss answered F. A. Taurinus, who had attempted to investigate the theory of parallels:

> In regard to your attempt, I have nothing (or not much) to say except that it is incomplete. It is true that your demonstration of the proof that the sum of the three angles of a plane triangle cannot be greater than 180° is somewhat lacking in geometrical rigor. But this in itself can easily be remedied, and there is no doubt that the impossibility can be proved most rigorously. But the situation is quite different in the second part, that the sum of the angles cannot be less than 180°; this is the critical point, the reef on which all the wrecks occur. I imagine that this problem has not engaged you very long. I have pondered it for over thirty years, and I do not believe that anyone can have given more thought to this second part than I, though I have never published anything on it.
>
> The assumption that the sum of the three angles is less than 180° leads to a curious geometry, quite different from ours [the Euclidean], but thoroughly consistent, which I have developed to my entire satisfaction, so that I can solve every problem in it with the exception of the determination of a constant, which cannot be designated *a priori*. The greater one takes this constant, the nearer one comes to Euclidean geometry, and when it is chosen infinitely large the two coincide. The theorems of this geometry appear to be paradoxical and, to the uninitiated, absurd; but calm, steady reflection reveals that they contain nothing at all impossible. For example, the three angles of a triangle become as small as one wishes, if only the sides are taken large enough; yet the area of the triangle can never exceed a definite limit, regardless of how great the sides are taken, nor indeed can it ever reach it.
>
> All my efforts to discover a contradiction, an inconsistency, in this non-Euclidean geometry have been without success, and the one thing in it which is opposed to our conceptions is that, if it were true, there must exist in space a linear magnitude, determined for itself (but unknown to us). But it seems to me that we know, despite the say-nothing word-wisdom of the metaphysicians, too little, or too nearly nothing at all, about the true nature of space, to

consider as *absolutely impossible* that which appears to us unnatural. If this non-Euclidean geometry were true, and it were possible to compare that constant with such magnitudes as we encounter in our measurements on the earth and in the heavens, it could then be determined *a posteriori.* Consequently, in jest I have sometimes expressed the wish that the Euclidean geometry were not true, since then we would have *a priori* an absolute standard of measure.

I do not fear that any man who has shown that he possesses a thoughtful mathematical mind will misunderstand what has been said above, but in any case consider it a private communication of which no public use or use leading in any way to publicity is to be made. Perhaps I shall myself, if I have at some future time more leisure than in my present circumstances, make public my investigations.

It is amazing that, despite his great reputation, Gauss was actually afraid to make public his discoveries in non-Euclidean geometry. He wrote to F. W. Bessel in 1829 that he feared "the howl from the Boeotians" if he were to publish his revolutionary discoveries.* He told H. C. Schumacher that he had "a great antipathy against being drawn into any sort of polemic."

The "metaphysicians" referred to by Gauss in his letter to Taurinus were followers of Immanuel Kant, the supreme European philosopher in the late eighteenth century and much of the nineteenth century. Gauss' discovery of non-Euclidean geometry refuted Kant's position that Euclidean space is *inherent in the structure of our mind.* In his *Critique of Pure Reason* (1781) Kant declared that "the concept of [Euclidean] space is by no means of empirical origin, but is an inevitable necessity of thought."

Another reason that Gauss withheld his discoveries was that he was a perfectionist, one who published only completed works of art. His devotion to perfected work was expressed by the motto on his seal, *pauca sed matura* ("few but ripe"). There is a story that the distinguished mathematician K. G. J. Jacobi often came to Gauss to relate new discoveries, only to have Gauss pull out some papers from his desk drawer

---

* An allusion to dull, obtuse individuals. .Actually, the "Boeotian" critics of non-Euclidean geometry—conceited people who claimed to have proved that Gauss, Riemann, and Helmholz were blockheads—did not show up before the middle of the 1870s. "If you witnessed the struggle against Einstein in the Twenties, you may have some idea of [the] amusing kind of literature [produced by these critics].... Frege, rebuking Hilbert like a schoolboy, also joined the Boeotians.... Your system of axioms, he said to Hilbert, is like a system of equations you cannot solve." (Freudenthal, 1962)

Carl Friedrich Gauss

that contained the very same discoveries. Perhaps it is because Gauss was so preoccupied with original work in many branches of mathematics, as well as in astronomy, geodesy, and physics (he coinvented an improved telegraph with W. Weber), that he did not have the opportunity to put his results on non-Euclidean geometry into polished form. These results were revealed among his private notes only after his death.

It would be impossible in this short book to present the tremendous range of Gauss' work. Gauss was called "the prince of mathematicians" and only Archimedes and Newton might be considered his equal. (For more detail on his life and work, see the biographies by Bell, Dunnington, and Hall.)

Lobachevsky

Another actor in this historical drama came along to steal the limelight from both J. Bolyai and Gauss: the Russian mathematician Nikolai Ivanovich Lobachevsky (1792–1856). He was the first to actually publish an account of non-Euclidean geometry (1829). His work attracted little attention when it appeared, largely because it was published in Russian and the Russians who read it were severely critical. In 1840 he published a treatise in German, which came to the attention of Gauss (who praised it in a letter to Schumacher, at the same time reiterating his own priority in the field).* At first Lobachevsky called his geometry "imaginary geometry," then later "pangeometry." In his published works he developed the subject quite thoroughly.

Lobachevsky openly challenged the Kantian doctrine of space as a subjective intuition. In 1835 he wrote: "The fruitlessness of the attempts made since Euclid's time... aroused in me the suspicion that the truth... was not contained in the data themselves; that to establish it the aid of experiment would be needed, for example, of astronomical observations, as in the case of other laws of nature."

Lobachevsky has been called "the great emancipator" by Eric Temple Bell; his name, said Bell, should be as familiar to every schoolboy as that of Michelangelo or Napoleon.† Unfortunately, Lobachevsky was not so

---

* See Bonola, 1955, for a translation of this paper.

† Bell, *The Search for Truth,* Chapter 14.

Nikolai Lobachevsky

appreciated in his lifetime; in fact, in 1846 he was fired from the University of Kazan, despite twenty years of outstanding service as a teacher and administrator. He had to dictate his last book in the year before his death, for by then he was blind.

It was not until after Gauss' death in 1855, when his correspondence was published, that the mathematical world began to take non-Euclidean ideas seriously. (Yet, as late as 1888 Lewis Carroll was poking fun at non-Euclidean geometry.) Some of the best mathematicians (Beltrami, Cayley, Klein, Poincaré, and Riemann) took up the subject, extending it, clarifying it, and applying it to other branches of mathematics, notably complex function theory. In 1868 the Italian mathematician Beltrami settled once and for all the question of a proof for the parallel postulate: he proved that no proof was possible! He did this by proving that non-Euclidean geometry is just as consistent as Euclidean geometry. (We will discuss his proof in the next chapter.)

Hyperbolic Geometry

Just what is non-Euclidean geometry? Technically speaking, it is any geometry different from Euclid's, and many such geometries are now known. We will restrict our attention to the particular geometry discovered by Gauss, J. Bolyai, and Lobachevsky, nowadays called *hyperbolic geometry* (see Appendix B for a discussion of elliptic geometry and other geometries discovered by Riemann). *Hyperbolic geometry is, by definition, the geometry you get by assuming all the axioms for neutral geometry and replacing Hilbert's parallel postulate by its negation, which we shall call the "hyperbolic axiom."*

HYPERBOLIC AXIOM.   In hyperbolic geometry there exist a line *l* and a point P not on *l* such that at least two distinct lines parallel to *l* pass through P [see Figure 6.1].

We can immediately see the main flaw in Legendre's attempted proof of the parallel postulate (pp. 19–21), namely, that the entire line *l* lies in

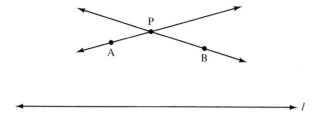

Figure 6.1

the interior of ⊀ APB without meeting either side, a phenomenon Legendre assumed to be impossible.

The following lemma (preliminary result) is the first important consequence of the hyperbolic axiom.

LEMMA 6.1   There exists a triangle whose angle sum is less than 180°.

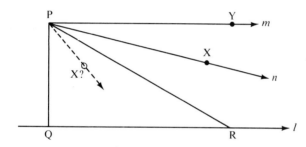

Figure 6.2

PROOF: Let l be a line and P a point not on l such that two parallels to l pass through P (hyperbolic axiom). We construct one parallel m as usual, by dropping perpendicular $\overleftrightarrow{PQ}$ to l and taking m to be the perpendicular to $\overleftrightarrow{PQ}$ through P. Let n be another parallel to l through P, $\overrightarrow{PX}$ a ray of n lying between $\overrightarrow{PQ}$ and a ray $\overrightarrow{PY}$ of m. By an application of Archimedes' axiom (Major Exercise 1), you can show that there is a point R on l, on the same side of $\overleftrightarrow{PQ}$ as X and Y, such that $(⊀ QRP)° < (⊀ XPY)°$. Moreover, $\overrightarrow{PR}$ lies in the interior of ⊀ QPX; otherwise, $\overrightarrow{PX}$ would have to lie in the interior of ⊀ QPR, hence, $\overrightarrow{PX}$ would meet QR (crossbar theorem, p. 69), contradicting our hypothesis than n does not meet l. Thus, $(⊀ RPQ)° < (⊀ XPQ)°$ by Theorem 4.3(3). Adding these inequalities gives $(⊀ QRP)° + (⊀ RPQ)° < 90°$, so that the angle sum of △QRP is less than 180°. ∎

Using this lemma, we can establish a universal version of the hyperbolic axiom. The parallel postulate in Euclidean geometry states that for every line and for every point off the line uniqueness of parallels holds. Its negation, the hyperbolic axiom, states that for *some* line *l* and *some* point P not on it uniqueness of parallels *fails* to hold. Could it be possible that in hyperbolic geometry uniqueness of parallels fails for some *l* and P but *holds for other l* and P? We will show that this is impossible.

UNIVERSAL HYPERBOLIC THEOREM.   In hyperbolic geometry, for every line *l* and every point P not on *l* there pass through P at least two distinct parallels to *l*.

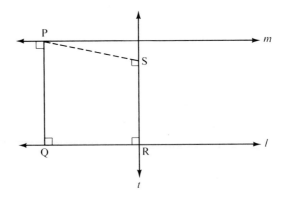

Figure 6.3

PROOF: Drop perpendicular $\overleftrightarrow{PQ}$ to *l* and erect line *m* through P perpendicular to $\overleftrightarrow{PQ}$. Let R be another point on *l*, erect perpendicular *t* to *l* through R, and drop perpendicular $\overleftrightarrow{PS}$ to *t*. (See Figure 6.3.) Now $\overleftrightarrow{PS}$ is parallel to *l* since they are both perpendicular to *t* (Corollary 1 to Theorem 4.1). We claim that *m* and $\overleftrightarrow{PS}$ are distinct lines. Assume on the contrary that S lies on *m*. Then □PQRS is a rectangle. But we showed in Theorem 4.7, that if one rectangle exists, then all triangles have angle sum 180°. This contradicts Lemma 6.1 proved above.■

Angle Sums (again)

The argument we have just given (using Lemma 6.1 and Theorem 4.7) proves the following theorem :

THEOREM 6.1. In hyperbolic geometry rectangles do not exist and all triangles have angle sum less than 180°.

If △ABC is any triangle, then 180° minus the angle sum of △ABC is a *positive* number. This number we called the *defect* of the triangle, and it plays a very important role in hyperbolic geometry (see Exercise 5 and Appendix A).

COROLLARY. In hyperbolic geometry all convex quadrilaterals have angle sum less than 360°.

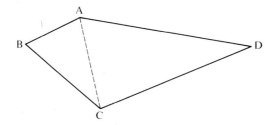

Figure 6.4

PROOF: Given any quadrilateral □ABCD. Take diagonal AC and consider triangles △ABC and △ACD; by the theorem, these triangles have angle sum <180°. The assumption that □ABCD is convex implies that $\overrightarrow{AC}$ is between $\overrightarrow{AB}$ and $\overrightarrow{AD}$, and $\overrightarrow{CA}$ is between $\overrightarrow{CB}$ and $\overrightarrow{CD}$, so that $(\angle BAC)° + (\angle CAD)° = (\angle BAD)°$ and $(\angle ACB)° + (\angle ACD)° = (\angle BCD)°$ (by Theorem 4.3(3)). By adding all six angles, we see that the angle sum of □ABCD is <360°. ∎

Similar Triangles

Next we shall consider Wallis' postulate, which cannot hold in hyperbolic geometry because, as we saw in Chapter 5 (pp. 125–127), it implies the parallel postulate. Thus, under certain circumstances similar triangles do not exist (negation of Wallis' postulate). But we can prove even more: under *no* circumstances do similar triangles exist! More precisely:

THEOREM 6.2.   In hyperbolic geometry if two triangles are similar, they are congruent. (In other words, AAA is a valid criterion for congruence of triangles.)

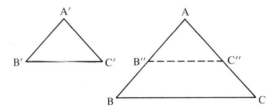

Figure 6.5

PROOF: Assume on the contrary that there exist triangles △ABC and △A′B′C′ which are similar but not congruent. Then no corresponding sides are congruent, otherwise the triangles would be congruent (ASA). Consider the triples (AB, AC, BC) and (A′B′, A′C′, B′C′) of sides of these triangles. One of these triples must contain at least two segments that are larger than the two corresponding segments of the other triple, e.g., AB > A′B′ and AC > A′C′. Then (by definition of >) there exist points B″ on AB and C″ on AC such that AB″ ≅ A′B′ and AC″ ≅ A′C′ (see Figure 6.5.). By SAS, △A′B′C′ ≅ △AB″C″. Hence, corresponding angles are congruent: ⊰ AB″C″ ≅ ⊰ B′, ⊰ AC″B″ ≅ ⊰ C′. By the hypothesis that △ABC and △A′B′C′ are similar, we also have ⊰ AB″C″ ≅ ⊰ B, ⊰ AC″B″ ≅ ⊰ C (Axiom C-5) This implies that $\overleftrightarrow{BC} \| \overleftrightarrow{B''C''}$ (Theorem 4.1 and Exercise 32, Chapter 4), so that quadrilateral ▱BB″C″C is convex. Also,   (⊰ B)° + (⊰ BB″C″)° = 180° = (⊰ C)° + (⊰ CC″B″)°   (Theorem 4.3(2) and 4.3(5)). It follows that quadrilateral ▱BB″C″C has angle sum 360°. This contradicts the Corollary to Theorem 6.1. ■

To sum up, in hyperbolic geometry it is impossible to magnify or shrink a triangle without distortion. In a hyperbolic world photography would be inherently surrealistic!

A startling consequence of Theorem 6.2 is that in hyperbolic geometry a segment can be determined with the aid of an angle, e.g., an angle of an equilateral triangle determines the length of a side uniquely. This is sometimes stated more dramatically by saying that hyperbolic geometry has an *absolute unit of length* (cf. Gauss' letter to Taurinus, p. 145). If the geometry of the physical universe were hyperbolic, it would no longer be necessary to keep a unit of length carefully guarded in the Bureau of Standards (the same is true for elliptic geometry).

Parallels That Admit a Common Perpendicular

In Chapter 5, commenting on the flaw in Proclus' attempted proof of the parallel postulate, I remarked that it was presumptuous to assume that parallel lines looked like railroad tracks, that is, that they were everywhere equidistant from each other. Let us now make this remark more precise. Given lines l and l' and points A, B, C . . . on l. Drop perpendiculars AA', BB', CC', . . . from these points to l'. We will say that points A, B, C, . . . are *equidistant* from l' if all these perpendicular segments are congruent to one another:

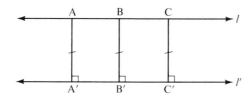

Figure 6.6

$$AA' \cong BB' \cong CC' \cong \ldots$$

THEOREM 6.3.   In hyperbolic geometry if l and l' are any distinct parallel lines, then any set of points on l equidistant from l' has at most two points in it.

PROOF: Assume on the contrary there is a set of three points A, B, and C on l equidistant from l'. Then quadrilaterals $\square$A'B'BA, $\square$A'C'CA, and $\square$B'C'CB are Saccheri quadrilaterals (the base angles are right angles and the sides are congruent). In Exercise 1, Chapter 5, you showed that the summit angles of a Saccheri quadrilateral are congruent. Thus, $\angle$ A'AB $\cong$ $\angle$ B'BA, $\angle$ A'AC $\cong$ $\angle$ C'CA, and $\angle$ B'BC $\cong$ $\angle$ C'CB. By transitivity (Axiom C-5), it follows that the supplementary angles $\angle$ B'BA and $\angle$ B'BC are congruent to each other; hence, by definition, they are right angles. Therefore, these Saccheri quadrilaterals are all rectangles. But rectangles do not exist in hyperbolic geometry (Theorem 6.1). This contradiction shows that A, B, and C cannot be equidistant from l'.∎

The theorem states that *at most* two points at a time on l can be equidistant from l'. It allows the possibility that there are pairs of points (A, B), (C, D), . . . on l such that each pair is equidistant from l'—thus, dropping perpendiculars, AA' $\cong$ BB' and CC' $\cong$ DD', but AA' is not congruent to CC'. A diagram for this might be Figure 6.7:

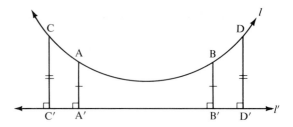

Figure 6.7

Figure 6.7 suggests that the point "in the middle" of *l* is closest to *l'*, with *l* moving away from *l'* symmetrically on either side of this middle point. We will prove that this is indeed the case (Theorems 6.4 and 6.5 and Exercises 4 and 10). Note, however, that Theorem 6.3 allows another possibility, that there is no pair of points on *l* equidistant from *l'*! A diagram for this might be Figure 6.8 :

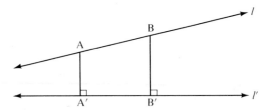

Figure 6.8
BB′ > AA′.

In Figure 6.8 the points on *l* are at varying distances from *l'*; *l* moves away from *l'* in one direction and approaches *l'* in the other direction without ever meeting it. Thus, different pairs of parallel lines need not resemble each other — some may look like the first diagram, some like the second.

THEOREM 6.4. In hyperbolic geometry if *l* and *l'* are parallel lines for which there exists a pair of points A and B on *l* equidistant from *l'*, then *l* and *l'* have a common perpendicular segment that is also the shortest segment between *l* and *l'*.

PROOF: Suppose A and B on *l* are equidistant from *l'*. Then □A′B′BA is a Saccheri quadrilateral, where A′ and B′ are the feet on *l'* of the perpendiculars from A and B. Let M be the midpoint of AB and M′ the midpoint of A′B′ (Proposition 4.3). The theorem will follow from the next lemma.

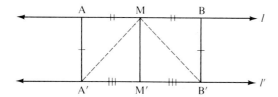

Figure 6.9

LEMMA 6.2.   The segment joining the midpoints of the base and summit of a Saccheri quadrilateral is perpendicular to both the base and the summit, and this segment is shorter than the sides.

PROOF: We know that ⊰ A ≅ ⊰ B (Exercise 1, Chapter 5). Hence, △A′AM ≅ △B BM (SAS). Therefore, the corresponding sides A′M and B′M are congruent. This implies △A′M′M ≅ △B′M′M (SSS). Therefore, the corresponding angles ⊰ A′M′M and ⊰ B′M′M are congruent. Since these are supplementary angles, they must be right angles, proving MM′ perpendicular to the base A′B′. From the two pairs of congruent triangles, we also have ⊰ A′MM′ ≅ ⊰ B′MM′ and ⊰ A′MA ≅ ⊰ B′MB. Adding the degrees of these angles, we have (⊰ AMM′)° = (⊰ BMM′)° (Theorem 4.3(3)), i.e., the supplementary angles ⊰ AMM′ and ⊰ BMM′ have the same number of degrees. Hence, they are right angles and MM′ is also perpendicular to the summit AB. Consider next quadrilateral □A′M′MA (Figure 6.10).

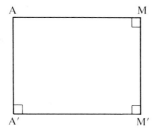

Figure 6.10

It has three right angles, so it is what we call a Lambert quadrilateral (Exercise 4, Chapter 5). In hyperbolic geometry the fourth angle must be acute, since rectangles do not exist (Theorem 6.1). You showed in Exercise 4(c), Chapter 5, that AA′ > MM′, i.e., that MM′ is shorter than AA′. The remainder of the proof that MM′ is shorter than any other segment between l and l' is left for Exercise 3. ∎

THEOREM 6.5.   In hyperbolic geometry if lines l and l' have a common perpendicular segment MM′, then they are parallel and MM′ is unique. Moreover, if A and B are any points on l such that M is the midpoint of segment AB, then A and B are equidistant from l'.

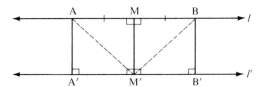

Figure 6.11

PROOF :   The fact that l and l' are parallel follows from the first corollary to the alternate interior angle theorem (Theorem 4.1). If l and l' had another common perpendicular segment NN′, then □M′N′NM would be a rectangle, which cannot exist (Theorem 6.1). Suppose now that M is the midpoint of AB. Drop perpendiculars AA′ and BB′ to l. We must prove that AA′ ≅ BB′. (See Figure 6.11.) First, △AM′M ≅ △BM′M (SAS), AM′ ≅ BM′ and ⊰ AM′M ≅ ⊰ BM′M. Therefore, (⊰ A′M′A)° = 90° − (⊰ AM′M)° = 90° − (⊰ BM′M)° = (⊰ B′M′B)° (by Theorem 4.3) so that ⊰ A′M′A ≅ ⊰ B′M′B. Hence, △AA′M′ ≅ △BB′M′ (AAS), so that the corresponding sides AA′ and BB′ are congruent.■

Limiting Parallel Rays

Theorems 6.4 and 6.5 and Exercises 4 and 10 give us a good under-standing of the first type of parallel lines. We know that such lines actually exist from the usual construction: start with any line l and any point P not on it (Figure 6.12). Drop perpendicular $\overrightarrow{PQ}$ to l and let m be the perpendicular through P to $\overrightarrow{PQ}$. Then m and l have the common perpendicular segment PQ. Pairs of points on m situated symmetrically about $\overrightarrow{PQ}$ are equidistant from l. By the universal hyperbolic theorem, there exist other lines n through P parallel to l. We cannot yet say that any such n is the second type of parallel, for n and l might have a common perpendicular going through a point other than P.

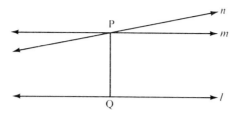

Figure 6.12

How then do we know that parallels of the second type exist? Here the axioms of continuity come in. The following is the intuitive idea (see Figure 6.13): Consider one ray $\overrightarrow{PS}$ of m, and consider various rays between $\overrightarrow{PS}$ and $\overrightarrow{PQ}$. Some of these rays, such as $\overrightarrow{PR}$, will intersect l, others such as $\overrightarrow{PY}$ will not. A continuity argument shows that as R recedes endlessly on l from Q, $\overrightarrow{PR}$ will approach a certain limiting ray $\overrightarrow{PX}$ that does *not* meet l. The ray $\overrightarrow{PX}$ is "limiting" in the following precise sense: any ray between $\overrightarrow{PX}$ and $\overrightarrow{PQ}$ intersects l, whereas any other ray $\overrightarrow{PY}$ such that $\overrightarrow{PX}$ is between $\overrightarrow{PY}$ and $\overrightarrow{PQ}$ does not intersect l. The ray $\overrightarrow{PX}$ may be called the *left limiting parallel ray* to l through P. Similarly, there is a *right limiting parallel ray* on the opposite side of $\overleftrightarrow{PQ}$.

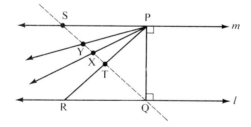

Figure 6.13

THEOREM 6.6.   For every line l and every point P not on l, let Q be the foot of the perpendicular from P to l. Then there are two unique rays $\overrightarrow{PX}$ and $\overrightarrow{PX'}$ on opposite sides of $\overleftrightarrow{PQ}$ that do not meet l and have the property that a ray emanating from P meets l if and only if it is between $\overrightarrow{PX}$ and $\overrightarrow{PX'}$. Moreover, these limiting rays are situated symmetrically about $\overleftrightarrow{PQ}$ in the sense that $\angle XPQ \cong \angle X'PQ$.

PROOF: To prove rigorously that $\overrightarrow{PX}$ exists, consider the line $\overleftrightarrow{SQ}$ (Figure 6.13). Let Σ_1 be the set of all points T on segment SQ, such that

$\overrightarrow{PT}$ meets l, together with all points on the ray opposite to $\overrightarrow{QS}$; let Σ_2 be the complement of Σ_1 (so $Q \varepsilon \Sigma_1$ and $S \varepsilon \Sigma_2$). By the crossbar theorem (p. 69), if point T on segment SQ belongs to Σ_1, then the entire segment TQ (in fact, $\overrightarrow{TQ}$) is contained in Σ_1. Hence, (Σ_1, Σ_2) is a Dedekind cut. By Dedekind's axiom (p. 80), there is a unique point X on $\overleftrightarrow{SQ}$ such that for P_1 and P_2 on $\overleftrightarrow{SQ}$, $P_1 * X * P_2$ if and only if $P_1 \varepsilon \Sigma_1$ and $P_2 \varepsilon \Sigma_2$.

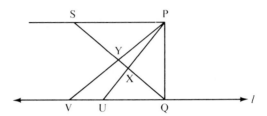

Figure 6.14

By definition of Σ_1 and Σ_2, rays below $\overrightarrow{PX}$ all meet l and rays above $\overrightarrow{PX}$ do not. We claim that $\overrightarrow{PX}$ does not meet l either. Assume on the contrary that $\overrightarrow{PX}$ meets l in a point U. Choose any point V on l to the left of U, i.e., $V * U * Q$ (Axiom B-2). Since V and U are on the same side of $\overleftrightarrow{SQ}$ (Exercise 9, Chapter 3), V and P are on opposite sides, so VP meets SQ in a point Y. We have $Y * X * Q$ (Proposition 3.7), so $Y \varepsilon \Sigma_2$, contradicting the fact that $\overrightarrow{PY}$ meets l. It follows that $\overrightarrow{PX}$ is the left limiting parallel ray (we obtain the right limiting parallel ray in a similar manner).

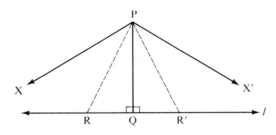

Figure 6.15

To prove symmetry, assume on the contrary that angles $\sphericalangle$ XPQ and $\sphericalangle$ X'PQ are not congruent, e.g., $(\sphericalangle XPQ)° < (\sphericalangle X'PQ)°$. By Axiom C-4, there is a ray between $\overrightarrow{PX'}$ and $\overrightarrow{PQ}$ that intersects l (by definition of limiting ray) in a point R' such that $\sphericalangle R'PQ \cong \sphericalangle XPQ$. Let R be the point on the opposite side of $\overrightarrow{PQ}$ from R' such that $R * Q * R'$ and

RQ ≅ R′Q (Axiom C-1). Then △RPQ ≅ △R′PQ (SAS). Hence, ⊰ RPQ ≅ ⊰ R′PQ, and by transitivity (Axiom C-5), ⊰ RPQ ≅ ⊰ XPQ. But this is impossible because $\overrightarrow{PR}$ is between $\overrightarrow{PX}$ and $\overrightarrow{PQ}$ (Axiom C-4).■

Either of the congruent angles ⊰ XPQ and ⊰ X′PQ is called (by an abuse of language) the *angle of parallelism* at point P with respect to *l*. Its degree measure is usually denoted Π(PQ)°. Note that Π(PQ)° < 90° for Π(PQ)° = 90° would contradict the universal hyperbolic theorem (see Exercise 7a). It can be shown that as P varies, Π(PQ)° takes on all possible values between 0° and 90° (see Major Exercise 9). One of the greatest discoveries by J. Bolyai and Lobachevsky is their formula for this number of degrees (see pp. 267–268). A *natural unit segment* OI in hyperbolic geometry is any segment OI such that Π(OI)° = 45°. Major Exercise 5 shows that all such segments are congruent to each other.

We have proved the existence of limiting parallel rays by a continuity argument. J. Bolyai actually discovered a simple method of *constructing* the limiting rays: Let Q be the foot of the perpendicular from P to *l*, *m* the line through P perpendicular to $\overleftrightarrow{PQ}$, R any point on *l* different from Q, and S the foot of the perpendicular from R to *m*. Then PR > QR (Exercise 3) and PS < QR (Exercise 4c, Chapter 5, on Lambert quadrilaterals). By the elementary continuity principle, a compass with center P and radius congruent to QR will intersect segment SR in a unique point X between S and R. It can be proved that $\overrightarrow{PX}$ is the right limiting parallel ray to *l* through P! (The proof is complicated: see pp. 224–225 or Bonola, pp. 216–221.)

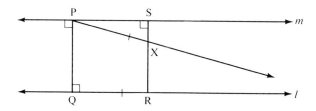

Figure 6.16

Classification of Parallels

We have discussed two types of parallels to a given *l*. The first type consists of parallels *m* such that *l* and *m* have a common perpendicular; *m* diverges from *l* on both sides of the common perpendicular. The second type consists of parallels that approach *l* asymptotically in one direction (i.e., they contain a limiting parallel ray in that direction) and which diverge from *l* in the other direction. If *m* is the second type of parallel, Exercises 7 and 8 show that *l* and *m* do not have a common perpendicular. We have implied that these two are the only types of parallels, and this is the content of the next theorem.

THEOREM 6.7.   Given *m* parallel to *l* such that *m* does not contain a limiting parallel ray to *l* in either direction. Then there exists a common perpendicular to *m* and *l* (which is unique by Theorem 6.5).

This theorem is proved by Borsuk and Szmielew (p. 291) by a continuity argument, but their proof gives you no idea of how to actually find the common perpendicular. There is an easy way to find it in the Klein and Poincaré models, discussed in the next chapter. Hilbert gave a direct construction, which we will sketch.

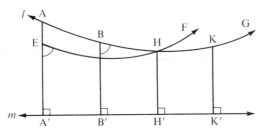

Figure 6.17

Hilbert's idea is to find two points H and K on *l* that are equidistant from *m*, for once these are found, the perpendicular bisector of segment HK is also perpendicular to *m* (see Lemma 6.2, p. 157). Choose any two points A and B on *l* and suppose that the perpendicular segment AA′ from A to *m* is longer than the perpendicular segment BB′ from B to *m*. (See Figure 6.17.) Let E be the point between A′ and A such that A′E ≅ B′B. On the same side of $\overleftrightarrow{AA'}$ as B, let $\overrightarrow{EF}$ be the unique ray such that ∢ A′EF ≅ ∢ B′BG, where A * B * G. The key point that will be proved in Major Exercises 2–6 is that $\overrightarrow{EF}$ intersects $\overrightarrow{AG}$ in a point H. Let K be the unique point on $\overrightarrow{BG}$ such that EH ≅ BK. Drop perpendiculars $\overleftrightarrow{HH'}$ and $\overleftrightarrow{KK'}$ to *m*. The upshot of these constructions is that □EHH′A′ is congruent to □BKK′B′ (just divide them into triangles). Hence, the corresponding sides HH′ and KK′ are congruent, so that the points H and K on *l* are equidistant from *m*, as required.∎

To sum up, given a point P not on *l*, there exist exactly two limiting parallel rays to *l* through P, one in each direction. There are infinitely many lines through P that do not enter the region between the limiting rays and *l*. Each such line is divergently parallel to *l* and admits a unique common perpendicular with *l* (for one of these lines the common perpendicular will go through P, but for all the rest the common perpendicular will pass through other points).

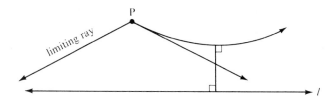

Figure 6.18

A note on terminology: In most books on hyperbolic geometry the word "parallel" is used only for lines that contain limiting parallel rays in our sense. The other lines, which admit a common perpendicular, have various names in the literature: "non-intersecting," "ultraparallel," "hyperparallel," and "superparallel." We will continue to use the word "parallel" to mean "non-intersecting." A parallel to *l* that contains a limiting parallel ray to *l* in a given direction will be called an *asymptotic parallel* in that direction, and a parallel to *l* that admits a common perpendicular to *l* will be called a *divergently parallel line*.

Strange New Universe?

In this chapter we have only begun to investigate the "strange new universe" of hyperbolic geometry. You can develop more of this geometry by doing the exercises, reading Appendix A, and examining works in the bibliography at the end of the book. You will encounter new entities such as asymptotic triangles, ideal and ultra-ideal points, equidistant curves, horocycles, and pseudospheres.

If you consider this geometry too "far out" to pursue, you are in for a surprise. We will see in the next chapter that if the undefined terms of hyperbolic geometry are suitably interpreted, hyperbolic geometry can be considered a part of Euclidean geometry!

Review Exercise

Which of the following statements are correct?

T (1) The negation of Hilbert's parallel postulate states that for every line *l* and every point P not on *l* there exist at least two lines through P parallel to *l*.

F (2) It is a theorem in neutral geometry that if lines *l* and *m* meet on a given side of a transversal *t*, then the sum of the degrees of the interior angles on that given side of *t* is less than 180°.

T (3) Gauss began working on non-Euclidean geometry when he was 15 years old.

T (4) The philosopher Kant taught that our minds could not conceive of any geometry other than Euclidean geometry.

T (5) The first mathematician to publish an account of hyperbolic geometry was the Russian Lobachevsky.

T (6) The crossbar theorem asserts that a ray emanating from a vertex A of △ABC and interior to ⅟A must intersect the opposite side BC of the triangle.

F (7) It is a theorem in hyperbolic geometry that for any segment AB there exists a square having AB as one of its sides.

T (8) Every Saccheri quadrilateral is a convex quadrilateral.

T (9) In hyperbolic geometry if △ABC and △DEF are equilateral triangles and ⅟A ≅ ⅟D, then the triangles are congruent.

F (10) In hyperbolic geometry, given a line *l* and a fixed segment AB, the set of all points on a given side of *l*, whose perpendicular segment to *l* is congruent to AB, equals the set of points on a line parallel to *l*.

F (11) In hyperbolic geometry any two parallel lines have a common perpendicular.

F (12) In hyperbolic geometry the fourth angle of a Lambert quadrilateral is obtuse.

F (13) In hyperbolic geometry some triangles have angle sum less than 180° and some triangles have angle sum equal to 180°.

T (14) In hyperbolic geometry if point P is not on line *l* and Q is the foot of the perpendicular from P to *l*, then the angle of parallelism for P with respect to *l* is the angle that a limiting parallel ray to *l* emanating from P makes with $\overrightarrow{PQ}$.

T (15) J. Bolyai showed how to construct limiting parallel rays using only the elementary continuity principle (p. 82) instead of Dedekind's axiom.

F (16) In hyperbolic geometry if *l*∥*m*, then there exist three points on *m* that are equidistant from *l*.

F (17) In hyperbolic geometry if *m* is any line parallel to *l*, then there exist two points on *m* which are equidistant from *l*.

F (18) In hyperbolic geometry if P is a point not lying on line *l*, then there are exactly two lines through P parallel to *l*.

F (19) In hyperbolic geometry if P is a point not lying on line *l*, then there are exactly two lines through P perpendicular to *l*.

F (20) In hyperbolic geometry if *l*‖*m* and *m*‖*n*, then *l*‖*n* (transitivity of parallelism).

F (21) In hyperbolic geometry if *m* contains a limiting parallel ray to *l*, then *l* and *m* have a common perpendicular.

T (22) In hyperbolic geometry if *l* and *m* have a common perpendicular, then there is one point on *m* that is closer to *l* than any other point on *m*.

T (23) In hyperbolic geometry if *m* does not contain a limiting parallel ray to *l*, and *m* and *l* have no common perpendicular, then *m* intersects *l*.

F (24) In hyperbolic geometry the summit angles of a Saccheri quadrilateral are right angles.

T (25) Every valid theorem of neutral geometry is also valid in hyperbolic geometry. T

F (26) In hyperbolic geometry opposite angles of any parallelogram are congruent to F *Lembut Qued* each other.

F (27) In hyperbolic geometry opposite sides of any parallelogram are congruent to F *Lenbert Qued* each other.

(28) In hyperbolic geometry, let ⊰ P be any acute angle, let X be any point on one side of this angle, and let Y be the foot of the perpendicular from X to the other side. If X recedes endlessly from P along its side, then Y will recede endlessly from P along its side.

F (29) In hyperbolic geometry if three points are not collinear, there is always a circle F that passes through them.

T (30) In hyperbolic geometry there exists an angle and there exists a line that lies entirely within the interior of this angle.

Exercises

The following are exercises in hyperbolic geometry. You are to assume the hyperbolic axiom and you can use the theorems presented in this chapter as well as any theorems of neutral geometry. However, do not use the Euclidean theorems stated in either Exercises 18–26, Chapter 5, or

Major Exercise 1, Chapter 5. (We can now assert that the theorems of neutral geometry are exactly those statements that are valid in *both* hyperbolic geometry and Euclidean geometry.)

1. Prove that if $\square$A'B'BA is a Saccheri quadrilateral ($\measuredangle$ A' and $\measuredangle$ B' are right angles and AA' $\cong$ BB'), then the summit AB is greater than the base A'B'. (Hint: join the midpoints M and M' and apply Exercise 4, Chapter 5, to the Lambert quadrilaterals $\square$A'M'MA and $\square$M'B'BM.)

2. Suppose that lines l and l' have a common perpendicular MM'. Let A and B be points on l such that M is *not* the midpoint of segment AB. Prove that A and B are not equidistant from l'.

3. Assume that the parallel lines l and l' have a common perpendicular segment MM'. Prove that MM' is the shortest segment between any point of l and any point of l'. (Hint: in showing MM' < AA', first dispose of the case in which AA' is not perpendicular to l' by means of Exercise 27, Chapter 4, and take care of the other case by Exercise 4, Chapter 5.)

4. Again, assume that MM' is the common perpendicular segment between l and l'. Let A and B be any points of l such that M * A * B, and drop perpendiculars AA' and BB' to l'. Prove that AA' < BB'. (Hint: use Exercises 2 and 4, Chapter 5; see Figure 6.19.)

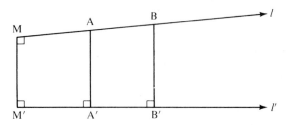

Figure 6.19

5. We have seen that in hyperbolic geometry the defect of any triangle is positive (Theorem 6.1). In Euclidean geometry all triangles have the same defect, namely, zero. In hyperbolic geometry could all triangles have the same defect? Assume that they do and use the additivity of the defect (Theorem 4.6) to derive a contradiction. Is there an upper bound for the defect of a triangle?

6. Given parallel lines l and m. Given points A and B that lie on the opposite side of m from l, i.e., for any point P on l, A and P are on opposite sides of m and B and P are on opposite sides of m. Prove that A and B lie on the same side of l.

7. (a) Prove that the angle of parallelism is acute by showing precisely how $\Pi(PQ)° = 90°$ implies that there is a unique parallel to l through P, contradicting the universal hyperbolic theorem.

(b) Let Let $\vec{PY}$ be a limiting parallel ray to l through P, and let X be a point on this ray between P and Y. It may seem intuitively obvious that $\vec{XY}$ is a limiting parallel ray to l *through X,* but this requires proof. Justify the steps that have not been justified.

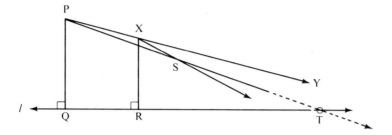

Figure 6.20

PROOF: (1) We must prove that any ray $\vec{XS}$ between $\vec{XY}$ and $\vec{XR}$ meets l, where R is the foot of the perpendicular from X to l. (2) S and Y are on the same side of $\overleftrightarrow{XR}$. (3) P and Y are on opposite sides of $\overleftrightarrow{XR}$. (4) By Exercise 6, S and Y are on the same side of $\overleftrightarrow{PQ}$. (5) S and R are on the same side of $\overleftrightarrow{XY} = \overleftrightarrow{PY}$. (6) Q and R are on the same side of $\overleftrightarrow{PY}$. (7) Q and S are on the same side of $\overleftrightarrow{PY}$. (8) Thus, $\vec{PS}$ lies between $\vec{PY}$ and $\vec{PQ}$, so it intersects l in a point T. (9) Point X is exterior to $\triangle PQT$. (10) $\overleftrightarrow{XS}$ does not intersect PQ. (11) Hence, $\overleftrightarrow{XS}$ intersects QT (Proposition 3.9a), so $\vec{XS}$ meets l.

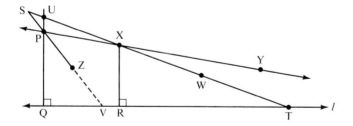

Figure 6.21

(c) Let us assume instead that $\vec{XY}$ is limiting parallel to l, with $P*X*Y$. Prove that $\vec{PY}$ is limiting parallel to l. (Hint: see Figure 6.21. You must show that $\vec{PZ}$ meets l in a point V. Choose any S such that $S*P*Z$. Show that SX meets $\vec{PQ}$ in a point U such that $U*P*Q$. Choose any W such that $U*X*W$, and show that $\vec{XW}$ is between $\vec{XY}$ and $\vec{XR}$ so that $\vec{XW}$ meets l in a point T. Apply Pasch's theorem to get V.)

8. Let $\overrightarrow{PX}$ be the right limiting parallel ray to l through P, and let Q and X' be the feet of the perpendiculars from P and X, respectively, to l. Prove that PQ > XX'. (Hint: use Exercise 7 to show that ⊰ X'XY is acute and that ⊰ X'XP is obtuse, so that Exercise 3, Chapter 5, can be applied to □PQX'X.) This exercise shows that the distance from X to l decreases as X recedes from P along a limiting parallel ray. In fact, one can prove that the distance from X to l approaches zero (see Major Exercise 11).

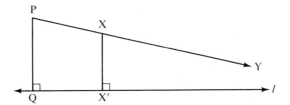

Figure 6.22

9. Let △ABC be any triangle, and let L, M, and N be the midpoints of BC, AB, and AC, respectively. Prove that △AMN is *not* similar to △ABC. Prove that MN is *not* congruent to BL by assuming the contrary and deducing that △ABC has angle sum 180°. (Hint: choose D such that M ∗ N ∗ D and ND ≅ MN. Show that △ANM ≅ △CND, then that △MDC ≅ △CBM. Substitute appropriately in the equation 180° = (⊰ BMC)° + (⊰ CMD)° + (⊰ AMN)° to get the result.)

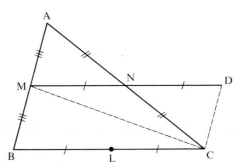

Figure 6.23

10. Assume that the parallel lines l and l' have a common perpendicular $\overleftrightarrow{PQ}$. For any point X on l, let X' be the foot of the perpendicular from X to l'. Prove that as X recedes endlessly from P on l, the segment XX' increases indefinitely. (Hint: we saw that it increases in Exercise 4. Drop a perpendicular XY to the limiting parallel ray between $\overrightarrow{PX}$ and $\overrightarrow{PX'}$. Use the crossbar theorem to show that $\overrightarrow{PY}$ intersects XX' in a point Z. Use Exercise 7 to show that P ∗ Y ∗ Z, hence that XX' > XY (why?). Conclude by applying Exercise 5, Chapter 5, to show that XY increases indefinitely as X recedes from P.)

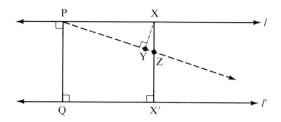

Figure 6.24

11. This problem has five parts. In the first part we will construct a Saccheri quadrilateral associated to any triangle; we will then apply this construction.

(a) Given $\triangle ABC$, let I, J, and K be the midpoints of BC, CA, and AB, respectively. Drop perpendiculars AD, BE, and CF from the vertices to $\overleftrightarrow{IJ}$. Prove that $AD \cong CF \cong BE$, and hence, that $\square EDAB$ is a Saccheri quadrilateral.

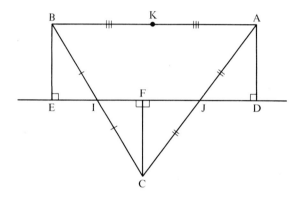

Figure 6.25

(b) Prove that the perpendicular bisector of AB, i.e., the perpendicular through K, is also perpendicular to $\overleftrightarrow{IJ}$, and hence, that $\overleftrightarrow{IJ}$ is divergently parallel to $\overleftrightarrow{AB}$.

(c) Recall that we denote the length of a segment by a bar, e.g., the length of MN is $\overline{MN}$ (Theorem 4.3B). Prove that $\overline{IJ} = \frac{1}{2}\overline{ED}$ (a separate argument is needed in case $\sphericalangle A$ or $\sphericalangle B$ is obtuse and the diagram is different). Deduce that in hyperbolic geometry $\overline{IJ} < \frac{1}{2}\overline{AB}$.

(d) Suppose now that $\sphericalangle C$ is a right angle. Prove that the Pythagorean theorem does not hold in hyperbolic geometry. (Hint: if the theorem were valid for right triangles $\triangle BCA$ and $\triangle ICJ$, then $\overline{IJ} = \frac{1}{2}\overline{AB}$ could be proved, contradicting part c of this exercise.)

(e) Suppose instead that $AC \cong BC$. Prove that K, F, and C are collinear but F is not the midpoint of CK (use Lemma 6.2 and part a of this exercise). For application of this result to mechanics, see Adler, pp. 192 and 253–257.

12. In Exercise 9, Chapter 5, we saw the elder Bolyai's false proof of the parallel postulate. The flaw in his argument was the assumption that *every* triangle has a circumscribed circle, i.e., that there is a circle passing through the three vertices of the triangle. The idea of the Euclidean proof of this assumption is to show that the perpendicular bisectors of the sides of the triangle meet in a point, and that this point is the center of the circumscribed circle. Figure out how Euclid's fifth postulate is used to prove that two of the perpendicular bisectors *l* and *m* have a common point (use Proposition 4.10) and then argue by congruent triangles to prove that the third perpendicular bisector passes through that point and that the point is equidistant from the three vertices. (Hint: join the common point D to the midpoint N of the third side, and prove that $\overleftrightarrow{DN}$ is perpendicular to the third side.)

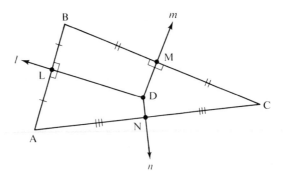

Figure 6.26

13. Part of the argument in Exercise 12 works for hyperbolic geometry, that is, *if* two of the perpendicular bisectors have a common point, then the third perpendicular bisector also passes through that point. In hyperbolic geometry there will be triangles for which two of the perpendicular bisectors are parallel (otherwise W. Bolyai's proof would be correct). Moreover, these perpendicular bisectors can be parallel in two different ways. Suppose that they are divergently parallel, that is, suppose that the perpendicular bisectors *l* and *m* have a common perpendicular *t* (see Figure 6.27). Prove that the third perpendicular bisector *n* is also perpendicular to *t*. (Hint: let A′, B′, and C′ be the feet on *t* of the perpendiculars dropped from A, B, and C, respectively. Let *l* bisect AB at L and be perpendicular to *t* at L′, and let *m* bisect BC at M and be perpendicular to *t* at M′. Let N be the midpoint of AC. Show by congruent triangles that AA′ ≅ BB′ and CC′ ≅ BB′ (draw AL′, BL′, BM′, and CM′). Hence, □C′A′AC is a Saccheri quadrilateral with N the midpoint of its summit AC. If N′ is the midpoint of the base A′C′, use Lemma 6.2 to show that $n = \overleftrightarrow{NN'}$ is perpendicular to *t* and $\overleftrightarrow{AC}$.) (See Major Exercise 7 for the asymptotically parallel case.)

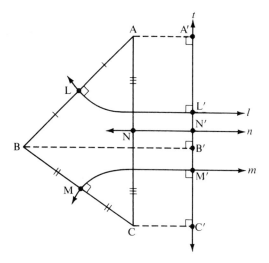

Figure 6.27

14. In Theorem 4.1 it was proved in neutral geometry that if alternate interior angles are congruent, then the lines are parallel. Strengthen this result in hyperbolic geometry by proving that the lines are divergently parallel, i.e., that they have a common perpendicular. (Hint: let M be the midpoint of transversal segment PQ and drop perpendiculars MN and ML to lines *m* and *l* (see Figure 6.28). Prove that L, M, and N are collinear by the method of congruent triangles.)

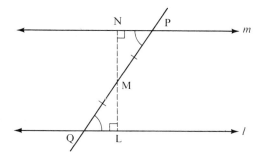

Figure 6.28

15. Show that any statement in the language of geometry that is a theorem in Euclidean geometry and whose negation is a theorem in hyperbolic geometry is equivalent to Hilbert's parallel postulate. Using this, make a long list of statements equivalent to Hilbert's parallel postulate, e.g., "the angle sum of every triangle is 180°." (In this exercise, "*P* equivalent to *Q*" means that $P \Leftrightarrow Q$ is a theorem in neutral geometry.)

16. Although the circumscribed circle may not exist for some triangles in hyperbolic geometry, prove that the inscribed circle always exists. (Hint: verify that the usual Euclidean proof—that the angle bisectors meet in a point equidistant from the sides—still works. Use the crossbar theorem.)

17. Comment on the following injunction by Saint Augustine: "The good Christian should beware of mathematicians and all those who make empty prophesies. The danger already exists that the mathematicians have made a covenant with the devil to darken the spirit and to confine man in the bonds of Hell."

Major Exercises

1. The purpose of this exercise is to fill the gap in the proof of Lemma 6.1 (p. 151). We must show that given any positive real number a, there is a point R on line l such that $(\sphericalangle QRP)^\circ < a^\circ$ (intuitively, by taking R sufficiently far out we can get as small an angle as we please). The idea is to construct a sequence of angles $\sphericalangle QR_1 P$, $\sphericalangle QR_2 P$, . . . each one of which is at most half the size of its predecessor. Justify the following steps:

Figure 6.29

There exists a point R_1 on l such that $PQ \cong QR_1$ (why?), so that $\triangle PQR_1$ is isosceles. It follows that $(\sphericalangle QR_1 P)^\circ \leq 45^\circ$ (why?). Next, there exists a point R_2 such that $Q * R_1 * R_2$ and $PR_1 \cong R_1 R_2$, so that $\triangle PR_1 R_2$ is isosceles. It follows that $(\sphericalangle QR_2 P)^\circ \leq 22\frac{1}{2}^\circ$ (to justify this step, use Corollary 1 to the Saccheri-Legendre theorem, p. 104). Continuing in this way, we get angles successively less than or equal to $11\frac{1}{4}^\circ$, $5\frac{5}{8}^\circ$, etc., so that by the Archimedean property of real numbers, we eventually get an angle $\sphericalangle QR_n P$ with $\sphericalangle (QR_n P)^\circ < a^\circ$.

2. *Symmetry of Limiting Parallelism.* We will prove that if $\overrightarrow{PB}$ is limiting parallel to $\overleftrightarrow{QD}$ (i.e., to $\overrightarrow{QD}$ in the direction of $\overrightarrow{QD}$), then $\overrightarrow{QD}$ is also limiting parallel to $\overrightarrow{PB}$. Justify the steps in the proof (see Figure 6.30).

PROOF: (1) The foot R of the perpendicular from Q to $\overleftrightarrow{PB}$ lies on $\overrightarrow{PB}$. (2) We must prove that every ray $\overrightarrow{QE}$ between $\overrightarrow{QR}$ and $\overrightarrow{QD}$ intersects $\overrightarrow{RB}$. (3) The foot F of the perpendicular from P to $\overleftrightarrow{QE}$ lies on $\overrightarrow{QE}$. (4) There is a unique point G between P and Q such that $PG \cong PF$. (5) Let $\overrightarrow{GH}$, on the same side of $\overleftrightarrow{PQ}$ as B, be perpendicular to $\overleftrightarrow{PQ}$. (6) There is a unique ray emanating from P on the same side of $\overleftrightarrow{PQ}$ as B that makes an angle with $\overleftrightarrow{PG}$ congruent to

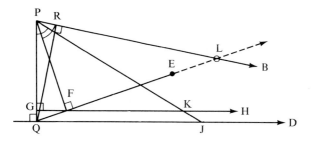

Figure 6.30

∢ FPB. (7) This ray intersects $\overrightarrow{QD}$ in a point J. (8) $\overrightarrow{GH}$ intersects side PQ of △PQJ but not side QJ, hence, it also intersects PJ in a point K. (9) Let L be the unique point on $\overrightarrow{PB}$ such that PL ≅ PK. (10) △PGK ≅ △PFL. (11) Hence, ∢ PFL is a right angle, so L must lie on $\overrightarrow{QE}$. ■

The idea behind this complicated argument is to rotate the figure consisting of rays $\overrightarrow{PB}$ and $\overrightarrow{FE}$, which we wish to meet, into the position of rays $\overrightarrow{PJ}$ and $\overrightarrow{GH}$, where we can show that the rays meet.

3. *Transitivity of Limiting Parallelism.* If $\overrightarrow{AB}$ and $\overrightarrow{CD}$ are both limiting parallel to $\overrightarrow{EF}$, then they are limiting parallel to each other. Justify the steps in the proof (Exercise 7 will be used).

PROOF: (1) $\overleftrightarrow{AB}$ and $\overleftrightarrow{CD}$ have no point in common. (2) Hence, there are two cases, depending on whether $\overleftrightarrow{EF}$ is between $\overleftrightarrow{AB}$ and $\overleftrightarrow{CD}$ or $\overleftrightarrow{AB}$ and $\overleftrightarrow{CD}$ are both on the same side of $\overleftrightarrow{EF}$. (3) In case $\overleftrightarrow{EF}$ is between $\overleftrightarrow{AB}$ and $\overleftrightarrow{CD}$, let G be the intersection of AC with $\overleftrightarrow{EF}$; we may assume G lies on ray $\overrightarrow{EF}$, otherwise we can consider $\overrightarrow{GF}$. (4) Any ray $\overrightarrow{AH}$ interior to ∢ GAB must intersect $\overrightarrow{EF}$ in a point I.

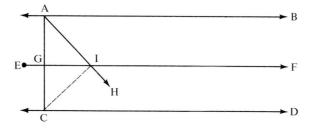

Figure 6.31

(5) $\overrightarrow{IH}$, lying interior to ∢ CIF, must intersect $\overrightarrow{CD}$. (6) Hence, any ray $\overrightarrow{AH}$ interior to ∢ CAB must intersect $\overrightarrow{CD}$, so $\overrightarrow{AB}$ is limiting parallel to $\overrightarrow{CD}$. (7) In case $\overleftrightarrow{AB}$ and $\overleftrightarrow{CD}$ are both on the same side of $\overleftrightarrow{EF}$, we may assume that $\overleftrightarrow{CD}$, e.g., is between $\overleftrightarrow{AB}$ and $\overleftrightarrow{EF}$ (see Figure 6.32). (8) Then AE intersects $\overleftrightarrow{CD}$ in a point G, which we may assume lies on ray $\overrightarrow{CD}$. (9) Any ray $\overrightarrow{AH}$ interior to ∢ GAB intersects $\overrightarrow{EF}$ in a point I. (10) Since $\overrightarrow{CD}$ enters △AEI at G and does not intersect side EI, it must intersect AI. (11) Therefore, $\overrightarrow{CD}$ is limiting parallel to $\overrightarrow{AB}$. ■

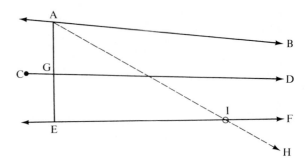

Figure 6.32

Note 1: The justification of Step 7 requires a lengthy argument; if you cannot supply it, see Moise, p. 316. Steps 8–11 prove the stronger result that any line between two asymptotically parallel lines is asymptotically parallel to both.

Note 2: If $\overrightarrow{PB}$ and $\overrightarrow{QD}$ are limiting parallel to each other, the convention is to pretend that they go through an "ideal point" Ω and denote them simply by $P\Omega$ and $Q\Omega$. The figure consisting of these rays and the segment PQ is then called a *singly asymptotic triangle* (in this definition, it is not assumed that PQ is necessarily perpendicular to $Q\Omega$). The next two exercises show that these triangles have some properties in common with ordinary triangles. (One can similarly define *doubly* (two ideal points) or *trebly* (three ideal points) *asymptotic triangles*.)

4. *Exterior Angle Theorem.* If $\triangle PQ\Omega$ is a singly asymptotic triangle, the exterior angles at P and Q are greater than their respective opposite interior angles. Justify the steps in the proof.

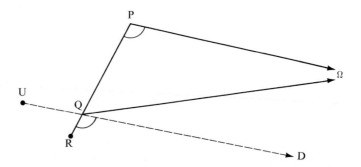

Figure 6.33

PROOF: (1) Given $R * Q * P$. We must show that $\measuredangle RQ\Omega$ is greater than $\measuredangle QP\Omega$. (2) Let $\overrightarrow{QD}$ be the unique ray on the same side of $\overleftrightarrow{PQ}$ as ray $Q\Omega$ such that $\measuredangle RQD \cong \measuredangle QP\Omega$. (3) If $U * Q * D$, then $\measuredangle UQP \cong \measuredangle QP\Omega$. (4) By Exercise 14, $\overleftrightarrow{QD}$ is divergently parallel to $\overleftrightarrow{P\Omega}$. (5) Hence, $\overrightarrow{QD}$ is between $\overrightarrow{QR}$ and $\overrightarrow{Q\Omega}$. (6) $\measuredangle RQ\Omega > \measuredangle QP\Omega$. ∎

5. *Congruence Theorem.* If in asymptotic triangles $\triangle AB\Omega$ and $\triangle A'B'\Omega'$ we have $\measuredangle BA\Omega \cong \measuredangle B'A'\Omega'$, then $\measuredangle AB\Omega \cong A'B'\Omega'$ if and only if $AB \cong A'B'$. Justify the steps in the proof and deduce as a corollary that $PQ \cong P'Q'$ if and only if $\Pi(PQ)° = \Pi(P'Q')°$.

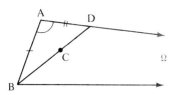
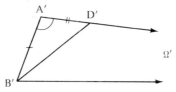

Figure 6.34

PROOF: (1) Assume $AB \cong A'B'$ and on the contrary $\measuredangle AB\Omega > \measuredangle A'B'\Omega'$. (2) There is a unique ray $\overrightarrow{BC}$ between $B\Omega$ and $\overrightarrow{BA}$ such that $\measuredangle ABC \cong \measuredangle A'B'\Omega'$. (3) $\overrightarrow{BC}$ intersects $A\Omega$ in a point D. (4) Let D' be the unique point on $A'\Omega'$ such that $AD \cong A'D'$. (5) Then $\triangle BAD \cong B'A'D'$. (6) Hence, $\measuredangle A'B'D' \cong \measuredangle A'B'\Omega'$,

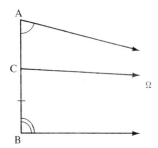
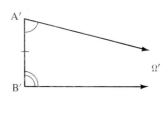

Figure 6.35

which is absurd. (7) Assume conversely that $\measuredangle AB\Omega \cong A'B'\Omega'$ and on the contrary $A'B' < AB$. (8) Let C be the point on AB such that $BC \cong B'A'$, and let $C\Omega$ be the ray from C limiting parallel to $A\Omega$. (9) Then $C\Omega$ is also limiting parallel to $B\Omega$. (10) By the first part of the proof, $\measuredangle BC\Omega \cong \measuredangle B'A'\Omega'$, hence, $\measuredangle BC\Omega \cong \measuredangle BA\Omega$. (11) But $\measuredangle BC\Omega > \measuredangle BA\Omega$, which is a contradiction. ■

6. *Conclusion of the Proof of Theorem* 6.7. We wish to show that $\overrightarrow{EF}$ intersects $\overrightarrow{AG}$ (see Figure 6.36). Justify the steps in the proof.

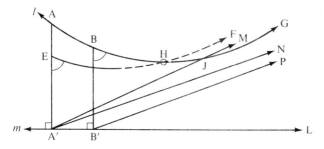

Figure 6.36

PROOF: Let $\overrightarrow{A'M}$ be limiting parallel to $\overleftrightarrow{EF}$, $\overrightarrow{A'N}$ limiting parallel to $\overrightarrow{AG}$, and $\overrightarrow{B'P}$ limiting parallel to $\overrightarrow{BG}$. (2) Since $EA' \cong BB'$ and $\measuredangle A'EF \cong \measuredangle B'BG$, we have $\measuredangle EA'M \cong \measuredangle BB'P$. (3) $\overrightarrow{B'L}$ differs from $\overrightarrow{B'P}$ and $\overrightarrow{A'L}$ differs from $\overrightarrow{A'N}$. (4) $\measuredangle MA'L \cong \measuredangle PB'L$. (5) $\overrightarrow{B'P}$ is limiting parallel to $\overrightarrow{A'N}$. (6) Hence, $\measuredangle NA'L$ is smaller than $\measuredangle PB'L$. (7) It follows that $\overrightarrow{A'M}$ lies between $\overrightarrow{A'N}$ and $\overrightarrow{A'A}$, so it must intersect $\overrightarrow{AG}$ in a point J. (8) J is on the same side of $\overleftrightarrow{EF}$ as A', hence, it is on the side opposite from A. (9) Thus, AJ intersects $\overleftrightarrow{EF}$ in a point H, which must be on $\overleftrightarrow{EF}$ because H is on the same side of $\overrightarrow{AA'}$ as J. ■

Where was the hypothesis of this theorem used?

7. In Exercises 12 and 13 we considered the perpendicular bisectors of the sides of $\triangle ABC$ and we showed that (1) if two of them have a common point, the third passes through that point; (2) if two of them have a common perpendicular, the third has that same perpendicular. It follows that if two of them are asymptotically parallel, then any two of them are asymptotically parallel. This result can be strengthened as follows: if perpendicular bisectors l and m are asymptotically parallel in the direction of ideal point Ω, then the third perpendicular bisector n is asymptotically parallel to l and m in the *same direction* Ω. Give the proof and justify each step. The proof is based on the following two lemmas:

Lemma 6.3   Given $\triangle ABC$. Let l, m, and n be the perpendicular bisectors of sides AB, BC, and AC at their midpoints L, M, and N, respectively. Let $\overline{AC} \geq \overline{AB}$ and $\overline{AC} \geq \overline{BC}$ (AC is the longest side). Then l, m, and n all intersect AC.

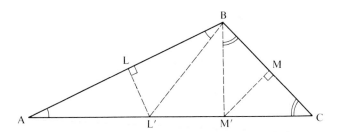

Figure 6.37

PROOF: (1) $(\measuredangle B)° \geq (\measuredangle A)°$ and $(\measuredangle B)° \geq (\measuredangle C)°$. (2) Hence, there is a point L' on AC such that $\measuredangle A \cong \measuredangle L'BA$, and a point M' on AC such that $\measuredangle C \cong \measuredangle M'BC$. (3) Then $AL' \cong BL'$ and $CM' \cong BM'$. (4) Thus, l is the line joining L to L' and $m = \overleftrightarrow{MM'}$. (5) It follows that all three perpendicular bisectors cut AC. ■

Lemma 6.4.   No line intersects all three sides of a trebly asymptotic triangle.

PROOF: (1) Suppose that a line t cuts l at Q and m at P. (2) Then ray $\overrightarrow{PQ}$ of t lies between the rays $P\Omega_2$ and $P\Omega_1$, which are limiting parallel to l. (3) $P\Omega_3$, the other ray through P that is limiting parallel to n, is opposite to $P\Omega_2$. (4) Hence,

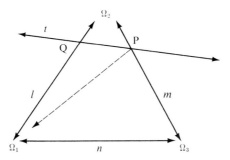

Figure 6.38

$P\Omega_1$ lies between $\overrightarrow{PQ}$ and $P\Omega_3$. (5) Thus, $\overrightarrow{PQ}$ does not intersect n. (6) Similarly, $\overrightarrow{QP}$ does not intersect n. ∎

8. Given any angle $\measuredangle\,A'OA$. It is a theorem in hyperbolic geometry that there is a unique line l called the *line of enclosure* of this angle such that l is limiting parallel to both sides $\overrightarrow{OA'}$ and $\overrightarrow{OA}$. Only the idea of the proof is given here; see if you can fill in the details (Wolfe, p. 97):

Assume that A and A′ are chosen so that $OA \cong OA'$. Let $A'\Omega$ be the limiting parallel ray to $\overrightarrow{OA}$ through A′, and $A\Sigma$ the limiting parallel ray to $\overrightarrow{OA'}$ through A. Let the rays r and r' be the bisectors of $\measuredangle\,\Sigma A\Omega$ and $\measuredangle\,\Omega A'\Sigma$, respectively. The idea of the proof is to show that the lines m and m' containing these rays are neither intersecting nor asymptotically parallel, so that, by Theorem 6.7, they have a unique common perpendicular l that turns out to be the line of enclosure of $\measuredangle\,A'OA$.

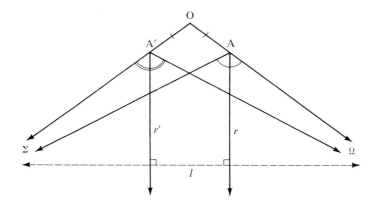

Figure 6.39

9. Use the result of the previous exercise to prove that every acute angle is an angle of parallelism, i.e., given an acute angle $\measuredangle\,BOA$ (Figure 6.40), there is a unique line l perpendicular to $\overleftrightarrow{BO}$ and limiting parallel to $\overrightarrow{OA}$. (Hint: reflect across $\overleftrightarrow{OB}$.)

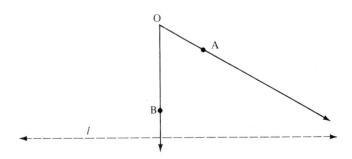

Figure 6.40

10. Let *l* and *m* be divergently parallel lines and let *t* be their common perpendicular cutting *l* at Q and *m* at P. Let *r* be a ray of *l* emanating from Q and *s* the ray of *m* emanating from P on the same side of *t* as *r*. Prove that there is a unique point R on *r* such that the perpendicular to *l* through R is limiting parallel to *s*. Prove also that for every point R' on *r* such that R'*R*Q, the perpendicular to *l* through R' is divergently parallel to *m*. (Hint: use Major Exercises 3 and 9.)

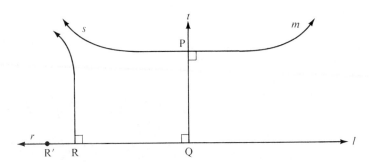

Figure 6.41

11. Let ray *r* emanating from point P be limiting parallel to line *l* and let Q be the foot of the perpendicular from P to *l*. Justify the terminology "asymptotically parallel" by proving that for any point R between P and Q there exists a point R' on ray *r* such that R'Q' ≅ RQ, where Q' is the foot of the perpendicular

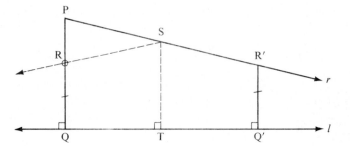

Figure 6.42

from R′ to *l*. (Hint: prove that the line through R that is asymptotically parallel to *l* in the opposite direction from *r* intersects *r* at a point S. Show that if T is the foot of the perpendicular from S to *l*, the point R′ obtained by reflecting R across line $\overleftrightarrow{ST}$ is the desired point.)

12. Let *l* and *n* be divergently parallel lines and PQ their common perpendicular segment. The midpoint S of PQ is called *the symmetry point* of *l* and *n*. Let *m* be the perpendicular to PQ through S. Let Ω and Ω′ be the ideal points of *l*, and let Σ and Σ′ be the ideal points of *n* (labeled as in Figure 6.43). By Major Exercise 8, there are unique lines "joining" these ideal points. Prove that (a) ΩΣ′ and ΣΩ′ meet at S; (b) *m* is perpendicular to both ΩΣ and Ω′Σ′. (Hint: use Major Exercise 5 and the symmetry part of Theorem 6.6.)

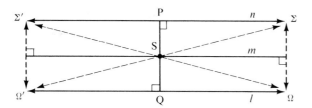

Figure 6.43

13. Here is another construction for the common perpendicular between divergently parallel lines *l* and *n*. It suffices to locate their symmetry point S, for a perpendicular can then be dropped from S to both lines. Take any segment AB on *l*. Construct point C on *l* such that B is the midpoint of AC and lay off any segment A′B′ on *n* congruent to AB. Let M, M′, N, and N′ be the midpoints of AA′, BB′, BA′, and CB′, respectively. Then the lines $\overleftrightarrow{MM'}$ and $\overleftrightarrow{NN'}$ are distinct and intersect at S. (If you can't prove this, see H. S. M. Coxeter's *Non-Euclidean Geometry,* 5th ed., 1968, p. 269, where it is deduced from Hjelmslev's midline theorem. Beware that Coxeter's description of midlines is partially wrong, e.g., no midline through S cuts *l* and *n*.)

14. M. Pieri has shown that the foundations of geometry can be built on the single undefined term "point" and the single undefined relation "point A is *equidistant* from points B and C." D. Scott has shown that, instead of the latter, one can use the single undefined relation "points A, B, C form a right triangle with right angle at A." It is obviously possible to define "A, B, C are collinear" in terms of "betweenness," namely, "A ∗ B ∗ C or B ∗ A ∗ C or A ∗ C ∗ B." What is not obvious is that in hyperbolic geometry it is possible to define "betweenness" in terms of "collinearity," as was done by F. P. Jenks, and that "collinearity" can in fact be taken as the single undefined relation for a hyperbolic geometry based on the elementary continuity principle (p. 82). Report on all of these results, using as a reference H. Royden's paper "Remarks on primitive notions for elementary Euclidean and non-Euclidean plane geometry" in *The Axiomatic Method,*

L. Henkin, P. Suppes and A. Tarski, eds., 1959, with corrections in W. Schwabhäuser's paper "Metamathematical Methods in Foundations of Geometry," in *Logic, Methodology and Philosophy of Science*, Y. Bar-Hillel, ed., Amsterdam: North Holland, 1965; and Blumenthal and Menger, 1970, p. 220.

15. Hilbert showed that all of plane hyperbolic geometry can be deduced from the incidence, betweenness, and congruence axioms, and an axiom asserting the existence of two non-opposite limiting parallel rays emanating from a given point not on a given line. Report on the proof of Saccheri's acute angle hypothesis from these axioms (see Wolfe, p. 78; Archimedes' axiom is not needed in this proof). Report also on the introduction of coordinates on the basis of these axioms (see W. Szmielew, "A new analytic approach to hyperbolic geometry," *Fundamenta Mathematicae*, **50** (1961): 129–158), and the use of such coordinates to prove the circular continuity principle (see J. Strommer, "Ein elementar Beweis des Kreisaxiome der hyperbolischen geometrie," *Acta Scientiarum Mathematicarum Szeged*, **22** (1961): 190–195).

16. If Dedekind's axiom is dropped from our axioms for hyperbolic geometry, then it is impossible to prove the existence of limiting parallel rays, for W. Pejas has constructed a "semi-elliptic" Archimedean geometry in which the hyperbolic axiom holds but any pair of parallel lines have a unique common perpendicular (see *Mathematische Annalen,* **143** (1961): 233). If Dedekind's axiom is replaced with the elementary continuity principle, then a proof of the existence of limiting parallel rays has been given by embedding in a metric projective plane (see Hessenberg and Diller, p. 239). Report on these results. If you could apply János Bolyai's construction (p. 161) for a more direct proof, you would probably be awarded a Ph.D.

Independence of the Parallel Postulate

7

> All my efforts to discover a contradiction, an inconsistency, in this non-Euclidean
> geometry have been without success....
>
> *C. F. Gauss*

Consistency of Hyperbolic Geometry

In the previous chapter you were introduced to hyperbolic geometry and presented with some theorems that must seem very strange to someone accustomed to Euclidean geometry. Even though you may admit that the proofs of these theorems are correct, given our assumptions, you may feel that the basic assumption of hyperbolic geometry—the hyperbolic axiom—is a false assumption. Let's examine what might be meant by saying it's false.

Suppose I assume that when I drop some object, say a stone, it will "fall" upward. I can go out and drop rocks and, unless I have rocks in my head, I will discover that my assumption was false.

Now what sort of experiment could I perform to show that the hyperbolic assumption is false, or, equivalently, to show that its negation, the parallel postulate, is true? First of all, I would have to understand what this statement means. In the above example I understood very well the meaning of "stone" and what it means to "drop" one, so I could act upon this understanding. But what does it mean that l is a "line," that P is a "point" not "on" l, or that there is a "unique parallel" to l through P? I might represent "points" and "lines" with paper, pencil, and straightedge. Suppose I draw $\overleftrightarrow{PQ}$ perpendicular to l and m through P perpendicular to $\overleftrightarrow{PQ}$, and then draw a line n through P, making a very small angle of $\varepsilon°$ with m. Using Euclidean trigonometry, I can calculate exactly how far out on n I would have to go to get to the point where n is supposed to intersect l, but if ε is small enough, that point might be millions of miles away. Thus, I could not physically perform the experiment to prove that the hyperbolic axiom is false.

But is geometry about lines that we can draw? Applied geometry (engineering) is; but pure geometry is about ideal lines, which are concepts, not objects. The only experiments we can perform on these ideal

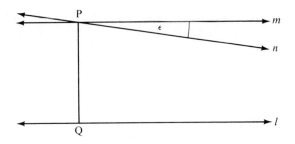

Figure 7.1

lines are thought experiments. So the question should be: can we *conceive* of a non-Euclidean geometry? Kant said no, that any geometry other than Euclidean is inconceivable. At the time, of course, no one had yet conceived of a different geometry. It is in this sense that Gauss, J. Bolyai, and Lobachevsky created a "new universe."

Other questions can be raised. Mathematicians reject many of their own ideas either because they lead to contradictions or do not lead anywhere, i.e., do not prove fruitful, useful, or interesting. Does the hyperbolic axiom lead to a contradiction? Saccheri thought it would, and tried to prove the parallel postulate that way. Is hyperbolic geometry fruitful, useful, or interesting?

Let us postpone this question and take up the former: Is hyperbolic geometry *consistent*? Nowadays we refer to this as a question in *metamathematics,* i.e., a question outside of a mathematical system about the system itself. The question is not about lines or points or other geometric entities; it is a question about the *whole system* of hyperbolic geometry.

If hyperbolic geometry were inconsistent, an ordinary mathematical argument could derive a contradiction. Saccheri tried to do this and failed. Could it be that he wasn't clever enough, that someday some genius will find a contradiction?

On the other hand, can it be proved that hyperbolic geometry is *consistent*—can it be proved that there is no possible way to derive a contradiction?

We might ask the same question about Euclidean geometry—how do we *know* it is consistent? Of course, this was never a burning question before the discovery of non-Euclidean geometry simply because everyone *believed* Euclidean geometry to be consistent. Remarkably enough, if we make this belief an explicit assumption (albeit a metamathematical assumption), it is possible to give a proof that hyperbolic geometry is consistent. Let us state this possibility as a theorem:

METAMATHEMATICAL THEOREM 1.   If Euclidean geometry is consistent, so is hyperbolic geometry.

Granting this result for the moment, we get the following important corollary.

COROLLARY.   If Euclidean geometry is consistent, then no proof or disproof of the parallel postulate from the rest of Hilbert's postulates will ever be found, i.e., the parallel postulate is independent of the other postulates.

To prove the corollary, assume on the contrary that a proof of the parallel postulate exists. Then hyperbolic geometry would be inconsistent, since the hyperbolic axiom contradicts a proved result. But Metamathematical Theorem 1 asserts that hyperbolic geometry is consistent relative to Euclidean geometry. This contradiction proves that no proof of the parallel postulate exists (RAA). The hypothesis that Euclidean geometry is consistent insures that no *dis*proof exists either.■

Thus, 2,000 years of efforts to prove Euclid V were in vain. There is no more hope of proving it than there is of finding a method for trisecting every angle using straight edge and compass alone.

Of course, when we say this, we are *assuming* the consistency of the venerable Euclidean geometry. Had Saccheri, Legendre, W. Bolyai, and others succeeded in proving Euclid V from the other axioms, with the noble intention of making Euclidean geometry more secure and elegant, they would have instead completely destroyed Euclidean geometry as a consistent body of thought! (I urge you, dear reader, to go over the preceding statements very carefully to make sure you have understood them. If you have not understood, you have missed the main point of this book.)

In the form given here, Metamathematical Theorem 1 is due to Eugenio Beltrami (1835–1900), although his proof was later simplified by Felix Klein (1849–1925).* However, these men were not the first to give plausible

---

* Beltrami made important contributions to differential geometry. Klein was a master of many branches of mathematics and an influential teacher. His book on the history of nineteenth-century mathematics shows how familiar he was with all aspects of the subject. Klein's famous inaugural address in 1872, his Erlanger *Programm,* made the study of groups of transformations and their invariants the key to geometry (see the "M-Exercises" at the end of this chapter).

arguments for the consistency of hyperbolic geometry. Taurinus, J. Bolyai, and Lobachevsky developed hyperbolic trigonometry out of ordinary trigonometry; if the latter were consistent, then the former also had to be consistent (the idea for this goes back to Lambert, who talked about trigonometry on "a sphere of imaginary radius"). Beltrami proved the relative consistency of hyperbolic geometry using differential geometry (the geometry of surfaces studied by means of the differential calculus, a subject developed by Gauss—see Appendix A). Klein recognized that projective geometry could be used to give another proof; he modified the method Arthur Cayley used for elliptic geometry.

To prove Metamathematical Theorem 1, we have to again ask ourselves, what is a "line" in hyperbolic geometry—in fact, what is the hyperbolic plane? The honest answer is that we don't know; it is just an abstraction. A hyperbolic "line" is an undefined term describing an abstract concept that resembles the concept of a Euclidean line except for its parallelism properties. Then how shall we visualize hyperbolic geometry? In mathematics, as in any other field of research, posing the right question is just as important as finding answers.

The question of "visualizing" means finding Euclidean objects that represent hyperbolic objects. This means finding a Euclidean *model* for hyperbolic geometry. In Chapter 2 (pp. 44–48) we discussed the idea of models for an axiom system; there we showed that the Euclidean parallel postulate is independent of the axioms for incidence geometry by exhibiting 3-point and 5-point models of incidence geometry that are not Euclidean. Here we want to know whether the parallel postulate is independent of a much *larger* system of axioms, namely neutral geometry. We can show that it is, and by the same method—by exhibiting models for hyperbolic geometry.*

The Beltrami-Klein Model

For brevity, we will refer to this first model as the "Klein model." We fix once and for all a circle γ in the Euclidean plane (which Cayley referred to as "the absolute"). If O is the center of γ and OR is a radius, the *interior* of γ by definition consists of all points X such that OX < OR.

---

* Unlike the situation for incidence geometry, we cannot construct a model for neutral geometry in which the elliptic parallel property holds because it is a theorem in neutral geometry that parallel lines exist (see Corollary 2 to Theorem 4.1).

Eugenio Beltrami

Felix Klein

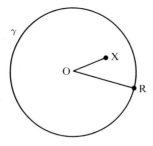

Figure 7.2

In Klein's model the points in the interior of γ represent the points of the hyperbolic plane.

Recall that a chord of γ is a segment AB joining two points A and B on γ. We wish to consider the segment without its endpoints, which we will call an *open chord* and denote by A)(B. In Klein's model the open chords of γ represent the lines of the hyperbolic plane. The relation "lies on" is represented in the usual sense: P lies on A)(B means that P lies on the Euclidean line $\overleftrightarrow{AB}$ and P is between A and B. The hyperbolic relation "between" is represented by the usual Euclidean relation "between." This much is easy. The representation of "congruence" is much more complicated, and we will discuss it later in this chapter (The Projective Nature of the Beltrami-Klein Model).

It is immediately clear from Figure 7.3 that the hyperbolic axiom holds in this representation:

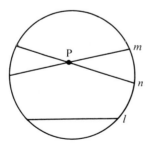

Figure 7.3

Here the two open chords m and n through P are both parallel to the open chord l—for what does "parallel" mean in this representation? The definition of "parallel" states that two lines are parallel if they have no point in common. In Klein's representation this becomes: two open chords are parallel if they have no point in common (in the definition of "parallel", replace the word "line" by "open chord"). The fact that the three chords, when extended, may meet outside the circle γ is irrelevant— points outside of γ do not represent points of the hyperbolic plane. So let us summarize the Beltrami-Klein proof of the relative consistency of hyperbolic geometry as follows:

First, a glossary is set up to "translate" the five undefined terms ("point," "line," "lies on," "between," and "congruent") into their interpretations in the Euclidean model (we have done this for the first four terms). All the defined terms are then interpreted by "translating" all occurrences of undefined terms. For instance, the defined term "parallel" was interpreted by replacing every occurrence of the word "line" in the definition by "open chord." Once all the defined terms have been interpreted, we have to interpret the axioms of the system. Incidence Axiom 1, for example, has the following interpretation in the Klein model:

INCIDENCE AXIOM 1 (KLEIN).   Given any two distinct points A and B in the interior of circle γ. There exists a unique open chord l of γ such that A and B both lie on l.

We must prove that this is a theorem in Euclidean geometry (and similarly, prove the interpretations of all the other axioms). Once all the interpreted axioms have been proved to be theorems in Euclidean geometry, any proof of a contradiction within hyperbolic geometry could be translated by our glossary into a proof of a contradiction in Euclidean geometry. From our assumption that Euclidean geometry is consistent, it follows that no such proof exists. Thus, if Euclidean geometry is consistent, so is hyperbolic geometry.

We must now backtrack and prove that the interpretations of the axioms of hyperbolic geometry in the Klein model are theorems in Euclidean geometry. Let us prove Axiom I-1 (Klein) stated above:

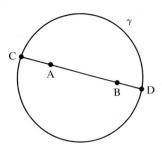

Figure 7.4

PROOF: Given A and B interior to γ. Let $\overleftrightarrow{AB}$ be the Euclidean line through them. This line intersects γ in two distinct points C and D. Then A and B lie on the open chord C)(D, and, by Axiom I-1 for Euclidean geometry, this is the only open chord on which they both lie.■

In the second step of the proof we used a theorem from Euclidean geometry that states that a line passing through the interior of a circle intersects the circle in two distinct points. This can be proved from the circular continuity principle (see Major Exercise 5, Chapter 3). Verifications of the interpretations of the other incidence axioms, the betweenness axioms, and Dedekind's axiom are left as exercises; the congruence axioms are verified later in the chapter (p. 216).

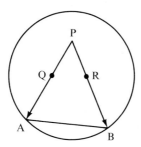

Figure 7.5

Limiting parallel rays.

One nice aspect of the Klein model is that it is easy to visualize the limiting parallel rays (see Figure 7.5). Let P be a point interior to γ and not on the open chord A)(B. A and B are points on the circle and therefore do not represent points in the hyperbolic plane; they are said to represent *ideal points* and are called the *ends* of the hyperbolic line represented by A)(B (see Note 2 following Major Exercise 3, Chapter 6). Then the limiting parallel rays to A)(B from P are represented by the segments PA and PB with the endpoints A and B omitted. It is clear that any ray between these limiting parallel rays intersects the open chord A)(B, whereas all other rays emanating from P do not. The symmetry and transitivity of limiting parallelism (which you saw in Major Exercises 2 and 3, Chapter 6, were tricky to prove) are utterly obvious in the Klein model, as is the fact that every angle has a *line of enclosure* (given ⊰ QPR, if A is the end of $\overrightarrow{PQ}$ and B is the end of $\overrightarrow{PR}$, then A)(B is the line of enclosure of ⊰ QPR—see Major Exercise 8, Chapter 6).

Let us conclude this section by considering the interpretation in the Klein model of "congruence," the subtlest part of the model. One method of interpretation is to use a system of numerical measurement of angle degrees and segment lengths. Two angles would then be interpreted as congruent if they had the same number of degrees, and two segments would be interpreted as congruent if they had the same length (cf. Theorem 4.3). The catch is that Euclidean methods of measuring degrees and lengths cannot be used. If we used Euclidean length, for example, then every line (i.e., open chord) would have a finite length less than or equal to the length of a diameter of γ. This would invalidate the interpretations of Axioms B-2 and C-1, which insure that lines are infinitely long.

We will further discuss the matter in this chapter (Perpendicularity in the Beltrami-Klein Model, p. 196 ff.; The Projective Nature of the Beltrami-Klein Model, p. 214 ff.), but first let's consider the Poincaré models, in which congruence of angles is easier to describe.

The Poincaré Models

A disk model due to Henri Poincaré (1854–1912)* also represents points of the hyperbolic plane by the points *interior* to a Euclidean circle γ, but lines are represented differently. First, all open chords that pass through the center O of γ (i.e., all *open diameters* l of γ) represent lines. The other lines are represented by *open arcs of circles orthogonal to* γ. More precisely, let δ be a circle orthogonal to γ (at each point of intersection of γ and

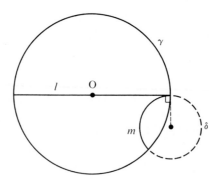

Figure 7.6

* Poincaré was the cousin of the president of France. Like Gauss, Poincaré made profound discoveries in many branches of mathematics and physics; he even started a new branch of mathematics, algebraic topology. He used his models of hyperbolic geometry to discover new theorems about automorphic functions of a complex variable. Poincaré is also important as a philosopher of science (see the next chapter).

δ the radii of γ and δ through that point are perpendicular). Then intersecting δ with the interior of γ gives an open arc m, which by definition represents a hyperbolic line in the Poincaré model. So we will call *Poincaré line,* or "P-line," either an open diameter l of γ or an open circular arc m orthogonal to γ (see Figure 7.6).

A point interior to γ "lies on" a Poincaré line if it lies on it in the Euclidean sense. Similarly, "between" has its usual Euclidean interpretation (for A, B, and C on an open arc coming from an orthogonal circle δ with center P, B is *between* A and C if $\overrightarrow{PB}$ is between $\overrightarrow{PA}$ and $\overrightarrow{PC}$).

A

Figure 7.7

The interpretation of *congruence for segments* in the Poincaré model is complicated, being based on a way of measuring length that is different from the usual Euclidean way, just as in the Klein model (see pp. 205–212). *Congruence for angles* has the usual Euclidean meaning, however, and this is the main advantage of the Poincaré model over the Klein model.* Remember that if two directed circular arcs intersect at a point A, the number of degrees in the *angle* they make is by definition the number of degrees in the angle between their tangent rays at A (see Figure 7.7). Or, if one directed circular arc intersects an ordinary ray at A, the number of degrees in the *angle* they make is by definition the number of degrees in the angle between the tangent ray and the ordinary ray at A.

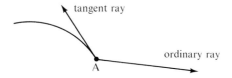

tangent ray

ordinary ray

A

Figure 7.8

---

* Technically, we say that the Poincaré model is *conformal*—it represents angles accurately—while the Klein model is not. Another example of a conformal model is Mercator's map of the surface of the earth.

Henri Poincaré

Having interpreted all the undefined terms of hyperbolic geometry in the Poincaré model, we get (by substitution) interpretations of all the defined terms. For example, two Poincaré lines are *parallel* if and only if they have no point in common. Then all the axioms of hyperbolic geometry get translated into statements in Euclidean geometry, and it will be shown in the section after next (Inversion in Circles) that these interpretations are theorems in Euclidean geometry. Hence, the Poincaré model furnishes another proof that if Euclidean geometry is consistent, so is hyperbolic geometry.

The limiting parallel rays in the Poincaré model are illustrated in Figure 7.9.

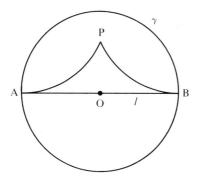

Figure 7.9

Here we have chosen l to be an open diameter A)(B; the rays are circular arcs that meet $\overleftrightarrow{AB}$ at A and B and are tangent to this line at those points. You can see how these rays approach l asymptotically as you move out toward the ideal points represented by A and B.

Figure 7.10 illustrates two parallel Poincaré lines with a common perpendicular. The diagram shows how m diverges from l on either side of the common perpendicular PO.

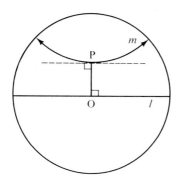

Figure 7.10

Figure 7.11 illustrates a Lambert quadrilateral. You can see that the fourth angle is acute.

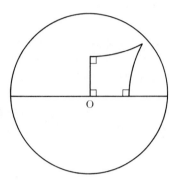

Figure 7.11

By adding the mirror image of this Lambert quadrilateral we get a diagram illustrating a Saccheri quadrilateral.

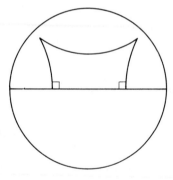

Figure 7.12

You may be surprised that we have two different models of hyperbolic geometry, one due to Klein and the other to Poincaré. (There is a third model, also due to Poincaré, soon to be described.) Yet you may have the feeling that these models are not "essentially different." In fact, these models are *isomorphic* in the technical sense that one-to-one correspondences can be set up between the "points" and "lines" in one model and the "points" and "lines" in the other so as to preserve the relations of incidence, betweenness, and congruence. Such isomorphism is illustrated in Figure 7.13. We start with the Klein model and consider, in Euclidean three-space, a sphere sitting on the plane of the Klein model and tangent to it at the origin. We project upward orthogonally the entire Klein model onto the lower hemisphere of this sphere; by this projection, the chords in the Klein model become arcs of circles orthogonal to the equator. We then project stereographically from the north pole of the sphere onto the original plane. The equator of the sphere will project onto a circle larger than the one used in the Klein model, and the lower hemisphere will project stereographically onto the inside of this circle. Under these successive transformations, the chords of the Klein model will be mapped one-to-one onto the diameters and orthogonal arcs of the Poincaré model. In this way the isomorphism of the models may be established.

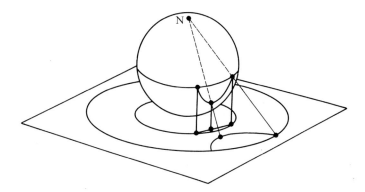

Figure 7.13

One can actually prove that *all possible models of hyperbolic geometry are isomorphic to one another,* i.e., that the axioms for hyperbolic geometry are *categorical.* The same is true for Euclidean geometry. The categorical nature of Euclidean geometry is established by introducing Cartesian coordinates into the Euclidean plane. Analogously, the categorical nature of hyperbolic geometry is established by introducing Beltrami coordinates into the hyperbolic plane (for which hyperbolic trigonometry must first be developed).*

In the other Poincaré model mentioned here, the points of the hyperbolic plane are represented by the points of one of the Euclidean half-planes determined by a fixed Euclidean line. If we use the Cartesian

* See Borsuk and Szmielew, Chapter 6.

model for the Euclidean plane, it is customary to make the x axis the fixed line and then to use for our model the upper half-plane consisting of all points (x, y) with $y > 0$. Hyperbolic lines are represented in two ways:

(1) as rays emanating from points on the x axis and perpendicular to the x axis;
(2) as semicircles in the upper half-plane whose center lies on the x axis (see Figure 7.14).

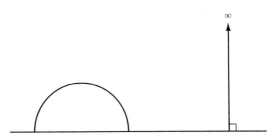

Figure 7.14

Incidence and betweenness have the usual Euclidean interpretation. This model is conformal also (degrees of angles are measured in the Euclidean way). Measurement of lengths will be discussed later.

To establish isomorphism with the previous models, choose a point E on the equator of the sphere in Figure 7.13, and let Π be the plane tangent to the sphere at the point diametrically opposite to E. Stereographic projection from E to Π maps the equator onto a line in Π and the lower hemisphere onto the lower half-plane determined by this line. Notice that the points on this line represent ideal points. However, one ideal point is missing: the point E got lost in the stereographic projection. It is customary to imagine an ideal "point-at-infinity" ∞ that corresponds to E; it is the common end of all the vertical rays.

Perpendicularity in the Beltrami-Klein Model

The Klein model is not conformal. Congruence of angles is interpreted differently from the usual Euclidean way, and will be explained later in this chapter (p. 216 ff.). Here we will describe only those angles that are congruent to their supplements, namely, right angles.

Let l and m be open chords of γ. To describe when $l \perp m$ in the Klein model, there are two cases to consider:

Case 1. One of l and m is a diameter. Then $l \perp m$ in the Klein sense if and only if $l \perp m$ in the Euclidean sense.

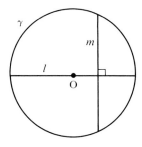

Figure 7.15

Case 2. Neither l nor m is a diameter. In this case we associate to l a certain point $P(l)$ outside of γ called the *pole* of l, defined as follows. Let t_1 and t_2 be the tangents to γ at the end points of l. Then by definition $P(l)$ is the unique point common to t_1 and t_2 (t_1 and t_2 are not parallel because l is not a diameter).

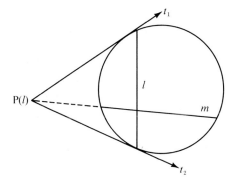

Figure 7.16

It turns out that *l is perpendicular to m in the sense of the Klein model if and only if the Euclidean line extending m passes through the pole of l.*

This description of perpendicularity will be justified later (p. 216 ff.). We can use it to see more easily why divergently parallel lines have a common perpendicular—Theorem 6.7. In the hypothesis of Theorem 6.7 we are given two parallel lines that do not contain limiting parallel rays.

In the Klein model this means that we are given open chords l and m that do not have a common end. The conclusion of Theorem 6.7 is that l and m have a common perpendicular k. How do we find k? Let's discuss Case 2, leaving Case 1 as an exercise. By the above description of perpendicularity if k were perpendicular to both l and m, the extension of k would have to pass through the pole of l and the pole of m. Hence, to construct k, we need only join these poles by a Euclidean line and take k to be the open chord of γ cut out by this line (Figure 7.17).*

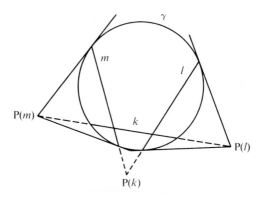

Figure 7.17

There is a nice language that describes the behavior of pairs of lines in the Klein model. Let us call the points inside circle γ (which represent all the points in the hyperbolic plane) *ordinary points*. We already called the points on the circle γ *ideal points*. Let us call the points outside γ *ultra-ideal points*. Finally, for every diameter of γ, let us imagine another point "at infinity" such that all the Euclidean lines parallel in the Euclidean sense to this diameter meet in this point at infinity, just as railroad tracks appear to meet at the horizon. These points at infinity will also be called *ultra-ideal*. We can then say that two Klein lines "meet" at an ordinary point, an ideal point, or an ultra-ideal point, depending on whether they are intersecting, asymptotically parallel, or divergently parallel, respectively. The ultra-ideal point at which divergently parallel Klein lines l and m "meet" is the pole $P(k)$ of their common perpendicular k (see Figure 7.17).

---

* If l and m did have a common end Ω, the Euclidean line joining $P(l)$ to $P(m)$ would be tangent to γ at Ω. That is why Saccheri claimed that asymptotically parallel lines have "a common perpendicular at infinity."

This language is suggestive of further theorems in hyperbolic geometry. For example, we know that two ordinary points determine a unique line, and we have seen that two ideal points also determine a unique line, the line of enclosure of Major Exercise 8, Chapter 6. We can ask the same question about two points that are ultra-ideal or about two points of different species. For example, an ordinary point and an ideal or ultra-ideal point always determine a unique line, but two ultra-ideal points may or may not (see Figure 7.18). Let us translate back from this language, say in the case of an ordinary point O and an ultra-ideal point P(*l*) that is the pole of a Klein line *l*. What is the Klein line "joining" O to P(*l*)? It is the unique Klein line *m* through O that is perpendicular in the sense of the Klein model to the line *l* (see Figure 7.16). We leave the other cases for exercises.

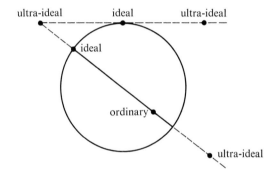

Figure 7.18

Inversion in Circles

In order to define congruence in the Poincaré models and verify the axioms of congruence, we must study the operation of inversion in a Euclidean circle; this operation will turn out to be the interpretation of reflection across a line in the hyperbolic plane. *This theory is part of Euclidean geometry,* so we may use the theorems you proved in Exercises 18–26, Chapter 5.

DEFINITION.   Let γ be a circle of radius r, center O. For any point $P \neq O$ the *inverse* P′ of P with respect to γ is the unique point P′ on ray $\overrightarrow{OP}$ such that $(\overline{OP})(\overline{OP'}) = r^2$ (where $\overline{OP}$ denotes the length of segment OP with respect to a fixed unit of measurement).

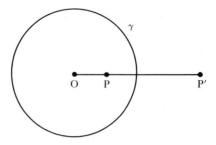

Figure 7.19

The following properties of inversion are immediate from the definition:

PROPOSITION 7.1. (a) P = P′ if and only if P lies on the circle of inversion γ. (b) If P is inside γ then P′ is outside γ, and if P is outside γ then P′ is inside γ. (c) (P′)′ = P.

The next two propositions tell how to construct the inverse point with a straightedge and compass.

PROPOSITION 7.2. Suppose P is inside γ. Let TU be the chord of γ through P which is perpendicular to $\overleftrightarrow{OP}$. Then the inverse P′ of P is the pole of chord TU, i.e., the point of intersection of the tangents to γ at T and U.

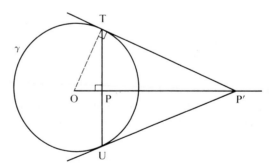

Figure 7.20

PROOF: Suppose the tangent to γ at T cuts $\overrightarrow{OP}$ at point P′. Right triangle $\triangle OPT$ is similar to right triangle $\triangle OTP′$ (since they have $\not\subset TOP$ in common and the angle sum is 180°). Hence, corresponding sides are proportional (Exercise 18, Chapter 5). As $\overline{OT} = r$, we get $(\overline{OP})/r = r/(\overline{OP'})$, which shows that P' is inverse to P. Reflecting across $\overleftrightarrow{OP}$ (Exercise 17, Chapter 4), we see that the tangent to γ at U also passes through P', so P' is indeed the pole of TU.■

PROPOSITION 7.3.   If P is outside γ, let Q be the midpoint of segment OP. Let σ be the circle with center Q and radius $\overline{OQ} = \overline{QP}$. Then σ cuts γ in two points T and U, $\overleftrightarrow{PT}$ and $\overleftrightarrow{PU}$ are tangent to γ, and the inverse P' of P is the intersection of TU and OP.

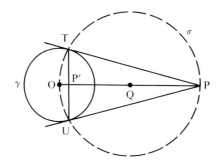

Figure 7.21

PROOF: By the circular continuity principle (p. 82), σ and γ do meet in two points T and U. Since $\angle$ OTP and $\angle$ OUP are inscribed in semicircles of σ, they are right angles (Exercise 24, Chapter 5), hence, $\overleftrightarrow{PT}$ and $\overleftrightarrow{PU}$ are tangent to γ. If TU meets OP in a point P', then P is the inverse of P' (Proposition 7.2), hence, P' is the inverse of P in γ.■

The next proposition shows how to construct the Poincaré line joining two ideal points – the line of enclosure. Its proof shows that $\angle$ OP'T in Figure 7.21 is indeed a right angle, as was needed in the proof of Proposition 7.3.

PROPOSITION 7.4.   Let T and U be points on γ that are not diametrically opposite and let P be the pole of TU. Then PT $\cong$ PU, $\angle$ PTU $\cong$ $\angle$ PUT, $\overleftrightarrow{OP} \perp \overleftrightarrow{TU}$, and the circle δ with center P and radius $\overline{PT} = \overline{PU}$ cuts γ orthogonally at T and U.

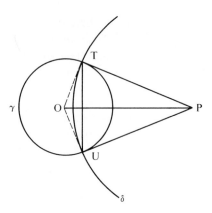

Figure 7.22

PROOF: By definition of pole, $\angle$ OTP and $\angle$ OUP are right angles, so by the hypotenuse-leg criterion, $\triangle$OTP $\cong$ $\triangle$OUP. Thus, PT $\cong$ PU, $\angle$ OPT $\cong$ $\angle$ OPU. The base angles $\angle$ PTU and $\angle$ PUT of the isosceles triangle $\triangle$TPU are then congruent, and the angle bisector $\overrightarrow{PO}$ is perpendicular to the base TU. The circle δ is then well-defined because $\overline{PT} = \overline{PU}$ and δ cuts γ orthogonally by our hypothesis that $\overleftrightarrow{PT}$ and $\overleftrightarrow{PU}$ are tangent to γ. ∎

LEMMA 7.1. Given that point O does not lie on circle δ. (a) If two lines through O intersect δ in pairs of points (P_1, P_2) and (Q_1, Q_2), respectively, then $(\overline{OP_1})(\overline{OP_2}) = (\overline{OQ_1})(\overline{OQ_2})$. This common product is called the *power* of O with respect to δ when O is outside δ, and minus this number is called the power of O when O is inside δ. (b) If O is outside δ and a tangent to δ from O touches δ at point T, then $(\overline{OT})^2$ equals the power of O with respect to δ.

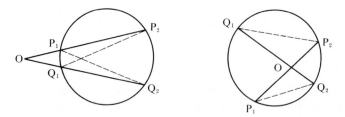

Figure 7.23

PROOF OF (a): Since angles that are inscribed in a circle and subtend the same arc are congruent (Exercise 25, Chapter 5), we have

$$\angle P_2 P_1 Q_2 \cong \angle P_2 Q_1 Q_2$$
$$\angle P_1 Q_2 Q_1 \cong \angle P_1 P_2 Q_1$$

(see Figure 7.23). It follows that $\triangle OP_1 Q_2$ and $\triangle OQ_1 P_2$ are similar, so that $(\overline{OP_1})/(\overline{OQ_1}) = (\overline{OQ_2})/(\overline{OP_2})$, as asserted.

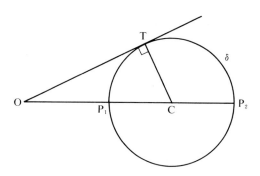

Figure 7.24

PROOF OF (b): Let C be the center of δ and let line OC cut δ at P_1 and P_2, with $O * P_1 * C * P_2$. By the Pythagorean theorem (Exercise 21, Chapter 5),

$$\begin{aligned}(\overline{OT})^2 &= (\overline{OC})^2 - (\overline{CT})^2 \\ &= (\overline{OC} - \overline{CT})(\overline{OC} + \overline{CT}) \\ &= (\overline{OC} - \overline{CP_1})(\overline{OC} + \overline{CP_2}) \\ &= (\overline{OP_1})(\overline{OP_2})\end{aligned}$$

(see Figure 7.24).■

PROPOSITION 7.5. Let P be any point which does not lie on circle γ and which does not coincide with the center O of γ, and let δ be a circle through P. Then δ cuts γ orthogonally if and only if δ passes through the inverse point P′ of P with respect to γ.

PROOF: Suppose first that δ passes through P′. Then the center C of δ lies on the perpendicular bisector of PP′ (Exercise 17, Chapter 4), hence, $\overline{CO} > \overline{CP}$ (Exercise 27, Chapter 4) and O lies outside δ. Therefore, there is a point T on δ such that the tangent to δ at T passes through O (Proposition 7.3). Lemma 7.1b then gives $(\overline{OT})^2 = (\overline{OP})(\overline{OP'}) = r^2$ so that T also lies on γ and δ cuts γ orthogonally.

Conversely, let δ cut γ orthogonally at points T and U. Then the tangents to δ at T and U meet at O, so that O lies outside δ. It follows that $\overrightarrow{OP}$ cuts δ again at a point Q. By Lemma 7.1(b), $r^2 = (\overline{OT})^2 = (\overline{OP})(\overline{OQ})$ so that Q = P', the inverse of P in γ. ∎

Proposition 7.5 can be used to construct the P-line joining two points P and Q inside γ that do not lie on a diameter of γ. First, construct the inverse point P', using Proposition 7.2. Then construct the circle δ determined by the three noncollinear points P, Q, and P' (use Exercise 12, Chapter 6). By Proposition 7.5, δ will be orthogonal to γ; intersecting δ with the interior of γ gives the desired P-line. This verifies the interpretation of Axiom I-1 for the Poincaré disk model. The verification is even simpler for the Poincaré upper half-plane model: Given two points P and Q that do not lie on a vertical ray, let the perpendicular bisector of Euclidean segment PQ cut the x axis at C. Then the semicircle centered at C and passing through P and Q is the desired P-line.

We could also have verified the interpretations of the incidence axioms, betweenness axioms, and Dedekind's axiom by using isomorphism with the Klein model (where the verifications are trivial).

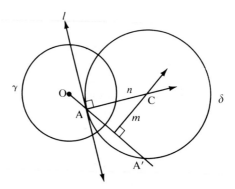

Figure 7.25

We turn now to the congruence axioms. Since angles are measured in the Euclidean sense in the Poincaré models, the interpretation of Axiom C-5 is trivially verified. Consider Axiom C-4, the laying off of a congruent copy of a given angle at some vertex A (for the disk model). If A is the center of γ, the angle is formed by diameters and the laying off is accomplished in the Euclidean way. If A is not the center O of γ, then the verification is a matter of finding a unique circle δ through A that is orthogonal to γ and tangent to a given Euclidean line l that passes through

A and not through O (since the tangents determine the angle measure). By Proposition 7.5, δ must pass through the inverse A' of A with respect to γ. The center C of δ must lie on the perpendicular bisector of chord AA' (Exercise 17, Chapter 4); call this bisector m. If δ is to be tangent to l at A, then C must also lie on the perpendicular n to l at A. So δ must be the circle whose center is the intersection C of m and n and whose radius is CA (see Figure 7.25).

To define congruence of segments in the disk model, we introduce the following definition of length:

DEFINITION. Let A and B be points inside γ, and let P and Q be the ends of the P-line through A and B. We define the *cross-ratio* (AB, PQ) by

$$(AB, PQ) = \frac{(\overline{AP})(\overline{BQ})}{(\overline{BP})(\overline{AQ})}$$

(where, e.g., $\overline{AP}$ is the Euclidean length of the Euclidean segment AP). We then define the *Poincaré length* $d(AB)$ by

$$d(AB) = \left|\log(AB, PQ)\right|.$$

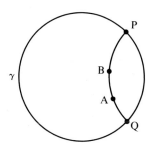

Figure 7.26

Notice first of all that this length does not depend on the order in which we write P and Q. For if (AB, PQ) = x, then (AB, QP) = $1/x$, and $\left|\log(1/x)\right| = \left|-\log x\right| = \left|\log x\right|$. Moreover, since (AB, PQ) = (BA, QP), we see that $d(AB)$ also does not depend on the order in which we write A and B.

We may therefore interpret the Poincaré segments AB and CD to be *Poincaré-congruent* if $d(AB) = d(CD)$. With this interpretation, Axiom C-2 is immediately verified.

Suppose we fix the point A on the P-line from P to Q and let point B move continuously from A to P, where $Q * A * B * P$, as in Figure 7.26. The cross-ratio (AB, PQ) will increase continuously from one to ∞, since $(\overline{AP})/(\overline{AQ})$ is constant, $\overline{BP}$ approaches zero, and $\overline{BQ}$ approaches $\overline{PQ}$. If we fix B and let A move continuously from B to Q, we get the same result. It follows immediately that for any Poincaré ray $\overrightarrow{CD}$, there is a unique point E on $\overrightarrow{CD}$ such that $d(CE) = d(AB)$, where A and B are given in advance. This verifies Axiom C-1.

We next verify Axiom C-3. This will follow immediately from the additivity of the Poincaré length, which asserts that if $A * B * C$ in the sense of the disk model, then $d(AC) + d(CB) = d(AB)$. To prove this additivity, label the ends so that $Q * A * B * P$. Then the cross-ratios (AB, PQ), (AC, PQ), and (CB, PQ) are all greater than one (because $\overline{AP} > \overline{BP}$, $\overline{BQ} > \overline{AQ}$, etc.); their logs are thus positive and we can drop the absolute value signs. We have

$$d(AC) + d(CB) = \log(AC, PQ) + \log(CB, PQ)$$
$$= \log[(AC, PQ)(CB, PQ)],$$

but $(AC, PQ)(CB, PQ) = (AB, PQ)$, as can be seen by cancelling terms.

Finally, to verify Axiom C-6 (SAS), we must study the effect of inversions on the objects and relations in the disk model.

DEFINITION. Let O be a point and k a positive number. The *dilation* with *center* O and *ratio* k is the transformation of the Euclidean plane that fixes O and maps a point $P \neq O$ onto the unique point P* on $\overrightarrow{OP}$ such that $\overline{OP^*} = k(\overline{OP})$ (so that points are moved radially from O a distance k times their original distance).

LEMMA 7.2. Let δ be a circle with center $C \neq O$ and radius s. Under the dilation with center O and ratio k, δ is mapped onto the circle δ^* with center C* and radius ks. If Q is a point on δ, the tangent to δ^* at Q* is parallel to the tangent to δ at Q.

PROOF: Choose rectangular coordinates so that O is the origin. Then the dilation is given by $(x, y) \to (kx, ky)$. The image of the line having equation $ax + by = c$ is the line having equation $ax + by = kc$, hence, the image is parallel to the original line. In particular, $\overleftrightarrow{CQ}$ is parallel to $\overleftrightarrow{C^*Q^*}$, and their perpendiculars at Q and Q*, respectively, are also parallel.

If δ has equation $(x - c_1)^2 + (y - c_2)^2 = s^2$, then δ^* has equation $(x - kc_1)^2 + (y - kc_2)^2 = (ks)^2$, from which the lemma follows. ∎

PROPOSITION 7.6.   Let γ be a circle of radius r and center O, δ a circle of radius s and center C. Assume that O lies outside δ; let p be the power of O with respect to δ (see Lemma 7.1). Let $k = r^2/p$. Then the image δ' of δ under inversion in γ is the circle of radius ks whose center is the image C^* of C under the dilation from O of ratio k. If P is any point on δ and P' is its inverse in γ, then the tangent t' to δ' at P' is the reflection across the perpendicular bisector of PP' of the tangent to δ at P.

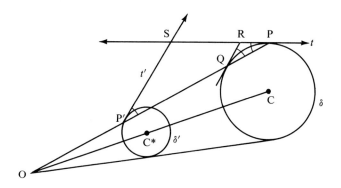

Figure 7.27

PROOF: Since O is outside δ, $\overrightarrow{OP}$ either cuts δ in another point Q or is tangent to δ at P (in which case let Q = P). Then

$$\frac{\overline{OP'}}{\overline{OQ}} = \frac{\overline{OP'}}{\overline{OQ}} \cdot \frac{\overline{OP}}{\overline{OP}} = \frac{r^2}{p},$$

which shows that P' is the image of Q under the dilation from O of ratio $k = r^2/p$. Hence, $\delta^* = \delta'$. By Lemma 7.2, the tangent t' to δ' at P' is parallel to the tangent u to δ at Q. Let t be tangent to δ at P. By Proposition 7.4, t and u meet at a point R such that $\angle RQP \cong \angle RPQ$. By Exercises 4, 5, and 32, Chapter 4, t and t' meet at a point S such that $\angle SP'P \cong \angle SPP'$. Since $\triangle PSP'$ is an isosceles triangle (base angles are congruent), S lies on the perpendicular bisector of PP'. Hence, t' is the reflection of t across this perpendicular bisector. ∎

COROLLARY.   Circle δ is orthogonal to circle γ if and only if δ is mapped onto itself by inversion in γ.

PROOF: If δ is orthogonal to γ and P lies on δ then $p = (\overline{OP})(\overline{OP'}) = r^2$ (Proposition 7.5 and Lemma 7.1), so $k = 1$ and $\delta = \delta'$. Conversely, if $\delta = \delta'$, then $p = r^2$, δ passes through the inverse P' of P in γ, so that by Proposition 7.5, δ is orthogonal to γ. ∎

LEMMA 7.3.   Let O be the center of circle γ, let P and Q be two points that are not collinear with O, and let P' and Q' be their inverses in γ. Then $\triangle$POQ is similar to $\triangle$Q'OP'.

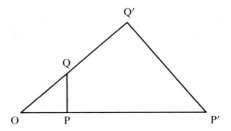

Figure 7.28

PROOF: The triangles have $\not\subset$ POQ in common and $(\overline{OP})(\overline{OP'}) = r^2 = (\overline{OQ})(\overline{OQ'})$. Thus, the SAS similarity criterion is satisfied (Exercise 20, Chapter 5). ∎

PROPOSITION 7.7.   Let l be a line not passing through the center O of circle γ. The image of l under inversion in γ is a punctured circle with missing point O. The diameter through O of the completed circle δ is (when extended) perpendicular to l.

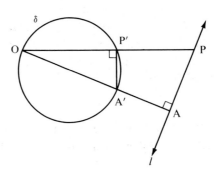

Figure 7.29

PROOF: Let A be the foot of the perpendicular from O to *l*, P be any other point on *l*, and A′ and P′ their inverses in γ. By Lemma 7.3, $\triangle$OP′A′ is similar to $\triangle$OAP. Hence, $\not\asymp$ OP′A′ is a right angle, so that P′ must lie on the circle δ having OA′ as diameter (Exercise 26, Chapter 5). Conversely, if we start with any point P′ on δ other than O and let $\overrightarrow{OP'}$ cut *l* in P, then reversing the above argument shows that P′ is the inverse of P in γ.■

PROPOSITION 7.8. Let δ be a circle passing through the center O of γ. The image of δ minus O under inversion in γ is a line *l* not through O; *l* is parallel to the tangent to δ at O.

PROOF: Let A′ be the point on δ diametrically opposite to O, let A be its inverse in γ, and *l* the line perpendicular to $\overleftrightarrow{OA}$ at A (see Figure 7.29). By the proof of Proposition 7.7, inversion in γ maps *l* onto δ minus O, hence, it must map δ minus O onto *l* (Proposition 7.1c). ■

It is obvious that reflection in a Euclidean line preserves the magnitude but reverses the sense of directed angles. The next proposition generalizes this to inversions.

PROPOSITION 7.9. A directed angle of intersection of two circles is preserved in magnitude but reversed in sense by an inversion. The same applies to the angle of intersection of a circle and a line or of two lines.

PROOF: Suppose that circles δ and σ intersect at point P with tangents *l* and *m* there. Let P′ be the inverse of P in γ, let δ' and σ' be the images of δ and σ under inversion in γ, and let *l*′ and *m*′ be their respective tangents at P′. The first assertion then follows from the fact that *l*′ and *m*′ are the reflections of *l* and *m* across the perpendicular bisector of PP′ (Proposition 7.6). The other cases follow from Propositions 7.7 and 7.8.■

The next proposition shows that inversion preserves the cross-ratio used to define Poincaré length.

PROPOSITION 7.10. If A, B, P, Q are four points distinct from the center O of γ and A′, B′, P′, Q′ are their inverses in γ, then (AB, PQ) = (A′B′, P′Q′).

PROOF: By Lemma 7.3, $(\overline{AP})/(\overline{OA}) = (\overline{A'P'})/(\overline{OP'})$ and $(\overline{AQ})/(\overline{OA}) = (\overline{A'Q'})/(\overline{OQ'})$, whence:

$$(1) \qquad \frac{\overline{AP}}{\overline{AQ}} = \frac{\overline{AP}}{\overline{OA}} \cdot \frac{\overline{OA}}{\overline{AQ}} = \frac{\overline{OQ'}}{\overline{OP'}} \cdot \frac{\overline{A'P'}}{\overline{A'Q'}}.$$

Similarly,

$$(2) \qquad \frac{\overline{BQ}}{\overline{BP}} = \frac{\overline{OP'}}{\overline{OQ'}} \cdot \frac{\overline{B'Q'}}{\overline{B'P'}}.$$

Multiplying equations (1) and (2) gives the result.■

PROPOSITION 7.11.   Let circle δ be orthogonal to circle γ. Then inversion in δ maps γ onto γ and maps the interior of γ onto itself. Inversion in δ preserves incidence, betweenness, and congruence in the sense of the Poincaré disk model inside γ.*

PROOF: The corollary to Proposition 7.6 tells us that γ is mapped onto itself. Suppose that P is inside γ and P′ is its inverse in δ. Let C be the center and s the radius of δ. Let $\overrightarrow{CP}$ cut γ at Q and Q′, so that by Proposition 7.5 $(\overline{CQ})(\overline{CQ'}) = s^2 = (\overline{CP})(\overline{CP'})$. Since P lies between Q and Q′, we have the inequalities CQ < CP < CQ′. Taking the reciprocal reverses inequalities, and we get $s^2/\overline{CQ} > s^2/\overline{CP} > s^2/\overline{CQ'}$, which is the same as CQ′ > CP′ > CQ. Thus, P′ lies between Q and Q′ and therefore is inside γ.

By Propositions 7.6, 7.8, and 7.9, inversion in δ maps any circle σ orthogonal to γ either onto another circle σ' orthogonal to γ or onto a line σ' orthogonal to γ, i.e., a line through the center O of γ. Obviously, the line σ joining O to C is mapped onto itself and any other line σ through O is mapped onto a circle σ' punctured at C, which is orthogonal to γ (by Propositions 7.7 and 7.9). In all these cases the above argument shows that the part of σ inside γ maps onto the part of σ' inside γ. Hence, P-lines are mapped onto P-lines.

If A and B are inside γ and P and Q are the ends of the P-line through

---

* It is easy to see that if, in the statement of Proposition 7.11 δ is taken to be a line through the center O of γ and "inversion" is replaced by "reflection," then the conclusion of Proposition 7.11 still holds.

A and B, then inversion in δ maps P and Q onto the ends of the P-line through A' and B'. By Proposition 7.10, $d(AB) = d(A'B')$, so congruence of segments is preserved. Proposition 7.9 shows that congruence of angles is also preserved. Furthermore, Poincaré betweenness is also preserved because B is between A and D if and only if A, B, and D are Poincaré collinear and $d(AD) = d(AB) + d(BD)$.■

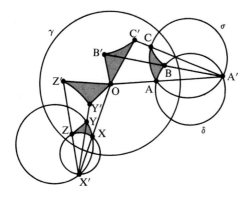

Figure 7.30

We come finally to the verification of the SAS axiom. We are given two Poincaré triangles $\triangle ABC$ and $\triangle XYZ$ inside γ such that ∢ A ≅ ∢ X, $d(AC) = d(XZ)$, and $d(AB) = d(XY)$. We must prove that the triangles are Poincaré-congruent. We first reduce to the case where $A = X = O$ (the center of γ): let δ be the circle orthogonal to γ through A and B, and σ the circle orthogonal to γ through A and C. Then δ again meets σ at point A' outside γ, which is inverse to A in γ (Proposition 7.5). Let ε be the circle centered at A' of radius s, where $s^2 = (\overline{AA'})(\overline{A'O})$. Since $\overline{AA'} = \overline{A'O} - \overline{AO}$, $s^2 = (\overline{A'O})^2 - (\overline{AO})(\overline{A'O}) = (\overline{A'O})^2 - r^2$, where r is the radius of γ. This equation shows that ε is orthogonal to γ (converse of the Pythagorean theorem). By definition of ε, O is the inverse of A in ε, and by Proposition 7.11, inversion in ε maps the Poincaré triangle $\triangle ABC$ onto a Poincaré-congruent Poincaré triangle $\triangle OB'C'$. In the same way, Poincaré triangle $\triangle XYZ$ can be mapped by inversion onto a Poincaré-congruent Poincaré triangle $\triangle OY'Z'$ (see Figure 7.30).

LEMMA 7.4. If $d(OB) = d$, then $\overline{OB} = r(e^d - 1)/(e^d + 1)$, where e is the base of the natural logarithm and r is the radius of γ.

PROOF: If P and Q are the ends of the diameter of γ through B, labeled so that $Q * O * B * P$, then $d = \log(OB, PQ)$. Exponentiating both sides of this equation gives

$$e^d = (OB, PQ) = \frac{\overline{OP}}{\overline{OQ}} \cdot \frac{\overline{BQ}}{\overline{BP}} = \frac{\overline{BQ}}{\overline{BP}} = \frac{r + \overline{OB}}{r - \overline{OB}}$$

and solving this equation for $\overline{OB}$ gives the result. ■

Returning to the proof of SAS, we have shown that we may assume that $A = X = O$. By Lemma 7.4 and the SAS hypothesis, we get $\overline{OB} = \overline{OY}$, $\overline{OC} = \overline{OZ}$, and $\angle BOC \cong \angle YOZ$. Hence, a suitable Euclidean rotation about O—combined, if necessary, with reflection in a diameter—will map Euclidean triangle $\triangle OBC$ onto Euclidean triangle $\triangle OYZ$. This transformation maps γ onto itself and the orthogonal circle through B and C onto the orthogonal circle through Y and Z, preserving Poincaré length and angle measure. Hence, the Poincaré triangles $\triangle OBC$ and $\triangle OYZ$ are Poincaré congruent. ■

This verification of SAS actually proves the following geometric description of Poincaré congruence:

THEOREM 7.1. Two triangles in the Poincaré disk model are Poincaré-congruent if and only if they can be mapped onto each other by a succession of inversions in circles orthogonal to γ and reflections in diameters of γ.

We will now apply the Poincaré model to determine the formula of J. Bolyai and Lobachevsky for the angle of parallelism (see p. 161). Let $\Pi(d)$ denote the number of *radians* in the angle of parallelism corresponding to the Poincaré distance d (the number of radians is $\pi/180$ times the number of degrees).

THEOREM 7.2. In the Poincaré disk model the formula for the angle of parallelism is $e^{-d} = \tan[\Pi(d)/2]$.

In this formula e is the base for the natural logarithm. The trigonometric tangent function is defined analytically as sin/cos, where the sine and cosine functions are defined by their Taylor series expansions (the tangent is *not* to be interpreted as the ratio of opposite to adjacent for a right triangle in the hyperbolic plane!).

PROOF: By definition of the angle of parallelism, d is the Poincaré distance $d(PQ)$ from some point P to some Poincaré line l, and $\Pi(d)$ is the number of radians in the angle that a limiting parallel ray to l through P makes with $\overrightarrow{PQ}$. We may choose l to be a diameter of γ and Q to be the center of γ, so that P lies on the perpendicular diameter. A limiting parallel ray through P is then an arc of a circle δ orthogonal to γ such that δ is tangent to l at one end Σ. The tangent line to δ at P therefore meets l at some interior point R that is the pole of chord $P\Sigma$ of

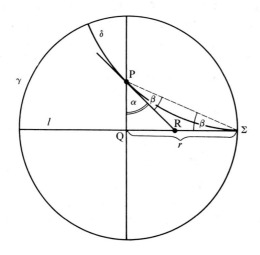

Figure 7.31

δ, and, by Proposition 7.4, $\measuredangle$ RPΣ and $\measuredangle$ RΣP both have the same number of radians β (see Figure 7.31). Let $\alpha = \Pi(d)$, which is the number of radians in $\measuredangle$ RPQ. Since 2β is the number of radians in $\measuredangle$ PRQ (exterior to $\triangle$PRΣ), we get $\alpha + 2\beta = \pi/2$, or $\beta = \pi/4 - \alpha/2$. The Euclidean distance $\overline{PQ}$ is $r \tan \beta$, so that, by the proof of Lemma 7.4,

$$e^d = \frac{1 + \tan \beta}{1 - \tan \beta}$$

Using the formula for β and the trigonometric identity

$$\tan (\pi/4 - \alpha/2) = \frac{1 - \tan \alpha/2}{1 + \tan \alpha/2},$$

we get the desired formula after some algebra. ■

We have developed only enough of the geometry of inversion in circles to verify the axioms in the Poincaré disk model. You will find further developments in the exercises. Inversion has many other applications in geometry, notably in Feuerbach's famous theorem on the nine-point circle of a triangle, the problem of Apollonius, and construction of linkages that change linear motion into curvilinear motion (see Eves, Kay, Kutuzov, and Pedoe; Eves uses the Poincaré model for a quick treatment of hyperbolic trigonometry).

The Projective Nature of the Beltrami-Klein Model

Having verified that the Poincaré disk interpretation is indeed a model of hyperbolic geometry, it follows from the isomorphism previously discussed (p. 195) that the Klein interpretation is also a model.

To be more explicit, consider the unit sphere Σ in Cartesian three-dimensional space given by the equation $x_1^2 + x_2^2 + x_3^2 = 1$. Let γ be the unit circle in the equatorial plane of Σ, determined by the equation $x_3 = 0$ and the equation for Σ. We will represent both the Poincaré disk and the Klein disk by the set Δ of points inside γ, and we will take as our isomorphism F the composite of two mappings: If N is the north pole $(0, 0, 1)$ of Σ, first project Δ onto the southern hemisphere of Σ stereographically from N. Then project orthogonally back upwards to the disk Δ (see Figure 7.32).

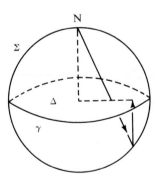

Figure 7.32

The isomorphism F will be considered to go from the Poincaré model to the Klein model. By an easy exercise in similar triangles, you can show that F is given in coordinates by

$$F(x_1, x_2, 0) = \left(\frac{2x_1}{1 + x_1^2 + x_2^2}, \frac{2x_2}{1 + x_1^2 + x_2^2}, 0 \right)$$

Or, if we ignore the third (zero) coordinate and use the single complex coordinate $z = x_1 + ix_2$, then F is given by

$$F(z) = \frac{2z}{1 + |z|^2}$$

It is clear that F maps the diameter of γ with ends P and Q onto the same diameter (but moving the points on the diameter out toward the circle). Let δ be a circle orthogonal to γ and cutting γ at points P and Q. We claim that F maps the Poincaré line with ends P and Q onto the open chord P)(Q. In fact, *if A is on the arc of δ from* P *to* Q *inside* γ, *then* $F(A)$ *is the point at which* $\overrightarrow{OA}$ *hits chord* PQ (see Figure 7.33).

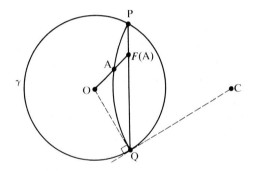

Figure 7.33

We can prove this as follows: Suppose the center C of δ has coordinates (c_1, c_2). By Proposition 7.3, the points P and Q are the intersections with γ of the circle having CO as diameter. After simplifying, the equation of this circle turns out to be

(1) $$x_1^2 - c_1 x_1 + x_2^2 - c_2 x_2 = 0.$$

Combining this equation with the equation $x_1^2 + x_2^2 = 1$ for γ gives the equation

(2) $$c_1 x_1 + c_2 x_2 = 1$$

for the line joining P to Q (called the *polar* of C with respect to γ). Since

δ is orthogonal to γ, $\measuredangle$ OQC is a right angle, and the Pythagorean theorem gives

(3) $$\overline{CQ}^2 = \overline{CO}^2 - \overline{OQ}^2 = c_1^2 + c_2^2 - 1$$

for the square of the radius of δ. Hence, δ is the circle

$$(x_1 - c_1)^2 + (x_2 - c_2)^2 = c_1^2 + c_2^2 - 1,$$

which simplifies to

(4) $$x_1^2 + x_2^2 = 2c_1 x_1 + 2c_2 x_2 - 1.$$

If now $A = (a_1, a_2)$ lies on δ, and $F(A) = (b_1, b_2)$ is its image under F, we have for $j = 1, 2$

(5) $$b_j = 2a_j/(1 + a_1^2 + a_2^2),$$

and combining this with equation 4 gives

(6) $$b_j = a_j/(c_1 a_1 + c_2 a_2).$$

It follows that

(7) $$c_1 b_1 + c_2 b_2 = 1,$$

hence, $F(A)$ lies on the polar of C, as asserted.■

We now use the isomorphism F to define congruence in the Klein model. Two segments (respectively, two angles) are interpreted to be *Klein-congruent* if their inverse images under F in the Poincaré model are Poincaré-congruent (as was defined before). With this interpretation, the verification of the congruence axioms is immediate. (It follows from this interpretation that the Klein model is conformal only at O.)

Next, let us justify the previous description of perpendicularity in the Klein model (p. 197). According to the above definition, two Klein lines l and m are Klein-perpendicular if and only if their inverse images $F^{-1}(l)$ and $F^{-1}(m)$ are perpendicular Poincaré lines. There are three cases to consider.

Case I. Both l and m are diameters. In this case it is clear that perpendicularity has its usual Euclidean meaning.

Case II. Only l is a diameter. Then $F^{-1}(l) = l$. The only way $F^{-1}(m)$, an arc of an orthogonal circle δ, can be perpendicular to l is if the Euclidean line extending l passes through the center C of δ (see Figure 7.34).

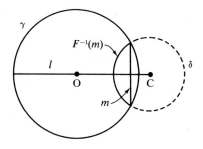

Figure 7.34

In that case the extension of l is the perpendicular bisector of chord m (Exercise 17, Chapter 4). Conversely, if l is perpendicular to m in the Euclidean sense, l bisects m, hence, the extension of l goes through C and l is then perpendicular to arc $F^{-1}(m)$.

Case III. Neither l nor m are diameters. Then $F^{-1}(l)$ and $F^{-1}(m)$ are arcs of circles δ and σ orthogonal to γ. Suppose δ is orthogonal to σ. By Proposition 7.4, the centers of these circles are the poles P(l) and P(m) of l and m, since these circles meet γ at the ends of l and m. Let P and Q be the ends of m. Inversion in δ interchanges P and Q, since this inversion maps both γ and σ onto themselves (Corollary to Proposition 7.6). But if P and Q are inverse in δ, the Euclidean line joining them has to pass through the center P(l) of δ (see Figure 7.35).

Conversely, if the extension of m passes through P(l), then P and Q are inverse to each other in δ (since points on γ are mapped onto γ by inversion in δ). By Proposition 7.5, σ is orthogonal to δ.∎

Next, let us describe the interpretation of reflections in the Klein model. In both Euclidean and hyperbolic geometries the *reflection* in a line m is

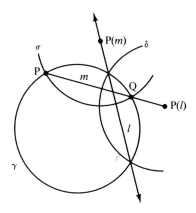

Figure 7.35

the transformation R_m of the plane, which leaves each point of m fixed and transforms a point A not on m as follows: Let M be the foot of the perpendicular from A to m. Then, by definition, $R_m(A)$ is the unique point A' such that A' * M * A and A'M $\cong$ MA. It is an easy exercise to show that reflection preserves incidence, betweenness, and congruence.

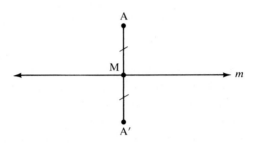

Figure 7.36

Returning to the Klein model, assume first that m is not a diameter of γ and let P be its pole. To drop a Klein-perpendicular from A to m, we draw the line joining A and P. Let it cut m at M and let t be the chord of γ cut out by this Euclidean line. Let Q be the pole of t and draw the line joining Q and A. Let this line cut γ at Σ and Σ' and let n be the open chord $\Sigma)(\Sigma'$. Draw the line joining Σ' and M and let it cut γ again at point Ω. If we now join Ω and Q, we obtain a line that cuts t at A' and γ again at Ω' (see Figure 7.37).

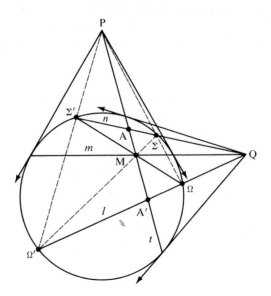

Figure 7.37

CONTENTION: The point A′ just constructed is the reflection in the Klein model of A across m. The Euclidean lines extending $\Omega\Sigma$ and $\Omega'\Sigma'$ meet at P and $\Omega\Sigma'$ meets $\Omega'\Sigma$ at point M.

One justification for this construction will be found in Major Exercise 12, Chapter 6. Here is another: Start with divergently parallel Klein lines $l = \Omega\Omega'$ and $n = \Sigma\Sigma'$ and their common perpendicular t. Let l meet t in A′ and n meet t in A, and let M be the midpoint of AA′ in the sense of the model. Let m be the Klein line through M Klein-perpendicular to t; m is obtained by joining M to the pole Q of t. Ray MΣ' is limiting parallel to n. If we reflect across m, then n is mapped onto the line through A′ Klein-perpendicular to t, namely, the line l. The end Σ' is mapped onto the end of l on the same side of t as Σ', namely, the point Ω'. Hence, ray MΣ' is mapped onto ray MΩ'. Now reflect across t; Ω' is sent to Ω, so MΩ' is mapped to MΩ. But successive reflections in the Klein-perpendicular lines m and t combine to give the 180° rotation about M. Hence, MΩ is the ray opposite to MΣ'. Similarly, MΣ is the ray opposite to MΩ'. Since reflection in m sent Σ' to Ω' and Σ to Ω, $\Sigma'\Omega'$ and $\Sigma\Omega$ must both be Klein-perpendicular to m and their Euclidean extensions meet at the pole P of m.

Secondly, let us describe the Klein reflection for the case in which m is a diameter of γ. In this case P is a point at infinity, t is perpendicular to m in the Euclidean sense, and M is the Euclidean midpoint of chord t (since a diameter perpendicular to a chord bisects it). Chord $\Omega\Sigma$ was shown to be perpendicular to diameter m in the argument above, so Ω is the Euclidean reflection of Σ across m. Hence, $\overset{\leftrightarrow}{Q\Omega}$ is the Euclidean reflection of $\overset{\leftrightarrow}{Q\Sigma}$ and we deduce that *A′ is the ordinary Euclidean reflection of A across diameter m* (see Figure 7.38).

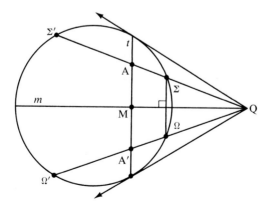

Figure 7.38

In order to describe the Klein reflection more succinctly, let us return to the notion of cross-ratio (AB, CD) defined by the formula

$$(AB, CD) = \frac{\overline{AC}}{\overline{AD}} \cdot \frac{\overline{BD}}{\overline{BC}}.$$

DEFINITION.   If A, B, C, and D are four distinct *collinear* points in the Euclidean plane such that $(AB, CD) = 1$, we say that C and D are *harmonic conjugates* with respect to AB and that ABCD is a *harmonic tetrad*. By symmetry of the cross-ratio, A and B are then also harmonic conjugates with respect to CD.

Another way to write the condition for a harmonic tetrad is $\overline{AC}/\overline{AD} = \overline{BC}/\overline{BD}$. Since C and D are distinct, one must be inside segment AB and the other outside (so that "C and D divide AB internally and externally in the same ratio"). Moreover, given AB, then C and D determine each other uniquely. For example, suppose $A * C * B$, and let $k = \overline{AC}/\overline{CB}$. If $k < 1$, then D is the unique point such that $D * A * B$ and $\overline{DB} = \overline{AB}/(1 - k)$, whereas if $k > 1$, then D is the unique point such that

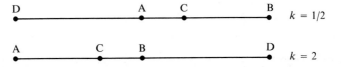

Figure 7.39

$A * B * D$ and $\overline{DB} = \overline{AB}/(k - 1)$. The case $k = 1$ is indeterminate, for there is no point D outside AB such that $\overline{AD} = \overline{BD}$. Thus, the midpoint M of AB has no harmonic conjugate. This exception can be removed by completing the Euclidean plane to the real projective plane by adding a "line at infinity" (see Major Exercise 4, Chapter 2). Then the harmonic conjugate of M is defined to be the "point at infinity" on $\overleftrightarrow{AB}$.

There is a nice way of constructing the harmonic conjugate of C with respect to AB with straightedge alone: Take any two points I and J collinear with C but not lying on $\overleftrightarrow{AB}$. Let $\overleftrightarrow{AJ}$ meet $\overleftrightarrow{BI}$ at point K and let $\overleftrightarrow{AI}$ meet $\overleftrightarrow{BJ}$ at point L. *Then $\overleftrightarrow{AB}$ meets $\overleftrightarrow{KL}$ at the harmonic conjugate D of C.*

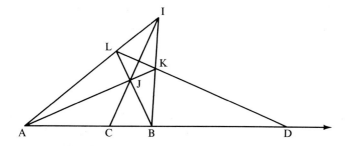

Figure 7.40

We will justify this *harmonic construction* on p. 222. Meanwhile, as a device to help remember the construction, "project" line $\overleftrightarrow{ID}$ to infinity. Then our figure becomes (Figure 7.41):

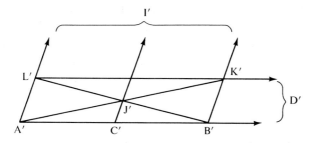

Figure 7.41

Since □A′B′K′L′ is now a parallelogram, we see that C′ is the midpoint of A′B′ and its harmonic conjugate is the "point at infinity" D′ on $\overrightarrow{A'B'}$. (This mnemonic device can be turned into a proof based on projective geometry—see Eves, Chapter 6.)

If you will now refer back to Figure 7.37, where the Klein reflection A′ of A was constructed, you will see that A′ *is the harmonic conjugate of* A *with respect to* MP. Just relabel the points in Figure 7.37 by the correspondences I–Σ′, J–Σ, K–Ω, L–Ω′, A–P, B–M, C–A, and D–A′ to obtain a figure for constructing the harmonic conjugate.

DEFINITION.   Let *m* be a line and P a point not on *m*. A transformation of the Euclidean plane called *the harmonic homology with center* P *and axis m* is defined as follows: Leave P and every point on *m* fixed. For any other point A let the line *t* joining

P to A meet m at M. Assign to A the unique point A' on t, which is the harmonic conjugate of A with respect to MP.

With this definition we can restate our result.

THEOREM 7.3.   Let m be a Klein line that is not a diameter of γ and let P be its pole. Then reflection across m is interpreted in the Klein model as restriction to the interior of γ of the harmonic homology with center P and with axis the Euclidean line extending m. If m is a diameter of γ, then reflection across m has its usual Euclidean meaning.

To justify the harmonic construction, we need the notion of a *perspectivity*. This is the mapping of a line l onto a line n obtained by projecting from a point P not on either line (Figure 7.42). It assigns to point A on l the point A' of intersection of $\overleftrightarrow{PA}$ with n. (Should $\overleftrightarrow{PA}$ be parallel to n, the image of A is the point at infinity on n.) P is called the *center* of this perspectivity.

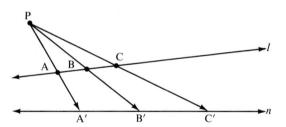

Figure 7.42

LEMMA 7.5.   A perspectivity preserves the cross-ratio of four collinear points, i.e., if A, B, C, and D are four points on line l and A', B', C', and D' are their images on line n under the perspectivity with center P, then $(AB, CD) = (A'B', C'D')$.

PROOF: By Exercise 23, Chapter 5, we have

$$\frac{\overline{AC}}{\overline{BC}} = \frac{\overline{AP}\sin \measuredangle APC}{\overline{BP}\sin \measuredangle BPC}$$

and

$$\frac{\overline{BD}}{\overline{AD}} = \frac{\overline{BP} \sin \, \sphericalangle BPD}{\overline{AP} \sin \, \sphericalangle APD}$$

which gives

$$(AB, CD) = \frac{(\sin \, \sphericalangle APC)(\sin \, \sphericalangle BPD)}{(\sin \, \sphericalangle BPC)(\sin \, \sphericalangle APD)}.$$

But $\sin \, \sphericalangle APC = \sin \, \sphericalangle A'PC'$, $\sin \, \sphericalangle BPD = \sin \, \sphericalangle B'PD'$, etc., so we obtain the same formula for $(A'B', C'D')$.∎

Now refer back to Figure 7.40. Let $\overleftrightarrow{IJ}$ meet $\overleftrightarrow{KL}$ at point M. Using the perspectivity with center I, Lemma 7.5 gives $(AB, CD) = (LK, MD)$, whereas using the perspectivity with center J we get $(AB, CD) = (KL, MD)$. But, by definition of cross-ratio, $(KL, MD) = 1/(LK, MD)$. Hence, (AB, CD) is its own reciprocal, which means $(AB, CD) = 1$, i.e., ABCD is a harmonic tetrad, as asserted. This justifies the harmonic construction on p. 221.∎

Next, we will apply Theorem 7.3 to calculate the length of a segment in the Klein model. According to our general procedure, length in the Klein model is defined by pulling back to the Poincaré model via the inverse of the isomorphism F and using the definition of length already given there. Thus, the length $d'(AB)$ of a segment in the Klein model is given by $d'(AB) = d(ZW) = |\log(ZW, PQ)|$, where $A = F(Z)$, $B = F(W)$, and P and Q are the ends of the Poincaré line through Z and W. By our earlier result illustrated in Figure 7.33, P and Q are also the ends of the Klein line through A and B.

The next theorem shows how to calculate $d'(AB)$ directly in terms of A, B, P, and Q. In its proof we will need the remark, "the cross-ratio (AB, PQ) is preserved by any Klein reflection." This is clear if we are reflecting in a diameter of γ. Otherwise, by Theorem 7.3, the Klein reflection is a harmonic homology whose center R lies outside γ. A reflection in the hyperbolic plane preserves collinearity, so for any Klein line l the mapping of l onto its Klein reflection n is just the perspectivity with center R. Therefore, Lemma 7.5 insures that the cross-ratio is preserved.

THEOREM 7.4.   If A and B are two points inside γ, and P and Q are the ends of the chord of γ through A and B, then the *Klein length* of segment AB is given by the formula

$$d'(AB) = \tfrac{1}{2}|\log(AB, PQ)|.$$

PROOF: We saw in the verification of the SAS axiom for the Poincaré disk model that any Poincaré line can be mapped onto a diameter by an inversion in a suitable orthogonal circle. Proposition 7.10 guarantees that cross-ratios are preserved by inversions. The transformation of the Klein model that corresponds to this inversion under our isomorphism F is a harmonic homology (Theorem 7.3), and this preserves cross-ratios of collinear points by the above remark. Hence, we may assume that A and B lie on a diameter.

Let $A = F(Z)$ and $B = F(W)$, so that, by definition, $d'(AB) = d(ZW)$. After a suitable rotation (which preserves cross-ratios), we may assume that the given diameter is the real axis. Its ends P and Q then have coordinates -1, $+1$. If Z and W have real coordinates z and w, then

$$(ZW, PQ) = \frac{1+z}{1-z} \cdot \frac{1-w}{1+w}$$

$$(AB, PQ) = \frac{1+F(z)}{1-F(z)} \cdot \frac{1-F(w)}{1+F(w)}.$$

But

$$1 - F(z) = 1 - \frac{2z}{1+|z|^2} = \frac{1 - 2z + |z|^2}{1+|z|^2}$$

$$1 + F(z) = \frac{1 + 2z + |z|^2}{1+|z|^2}$$

$$\frac{1 + F(z)}{1 - F(z)} = \frac{1 + 2z + |z|^2}{1 - 2z + |z|^2}.$$

Since z is real, $z = \pm|z|$ and we get

$$\frac{1 + F(z)}{1 - F(z)} = \left(\frac{1+z}{1-z}\right)^2.$$

From this and the formula obtained from it by substituting w for z, it follows that $(AB, PQ) = (ZW, PQ)^2$, and taking logarithms of both sides proves the theorem. ■

Finally, let us apply our results to justify J. Bolyai's construction of the limiting parallel ray (p. 161). We are given a Klein line l and a point P not on it. Point Q on l is the foot of the Klein-perpendicular t from P to l, and m is the Klein-perpendicular to t through P. Let R be any other point on l, and S the foot on m of the Klein-perpendicular from R. Bolyai's construction is based on the contention that if the limiting parallel ray to l from P in the direction $\overrightarrow{QR}$ meets RS at X, then PX is Klein congruent to QR.

Let T and M be the poles of t and m. Let Ω and Ω' be the ends of l. If we join these ends to M, the intersections Σ and Σ' with γ will be the ends of the Klein reflection n of l across m.

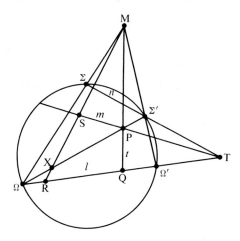

Figure 7.43

As Figure 7.43 shows, the collinear points Ω, X, P, and Σ' are in perspective with the collinear points Ω, R, Q, and Ω' (in that order), the center of the perspectivity being M. By Lemma 7.5, such a perspectivity preserves cross-ratios, so that $(XP,\Omega\Sigma') = (RQ,\Omega\Omega')$. Theorem 7.4 tells us that $d'(XP) = d'(RQ)$, justifying Bolyai's contention. (In case m is a diameter of γ, M is a point at infinity; then instead of Lemma 7.5 we use the parallel projection theorem (see p. 137) to deduce the above equality of cross-ratios.) ∎

Review Exercise

Which of the following statements are correct?

(1) Although two thousand years of efforts to prove the parallel postulate as a theorem in neutral geometry have been unsuccessful, it is still possible that someday some genius will succeed in proving it.

(2) If we add to the axioms of neutral geometry the elliptic parallel postulate (that no parallel lines exist), we get another consistent geometry called elliptic geometry.

(3) All the ultra-ideal points in the Klein model are points in the Euclidean plane outside γ.

(4) Both the Klein and Poincaré models are "conformal" in the sense that congruence of angles has the usual Euclidean meaning.

(5) In the Poincaré model "lines" are represented by all open diameters of a fixed circle γ and by all open arcs inside γ of circles intersecting γ.

(6) For any chord A)(B whatever of circle γ, the tangents to γ at the end points A and B of the chord meet in a unique point called the "pole" of that chord.

(7) In the Poincaré model two Poincaré lines are interpreted as "perpendicular" if and only if they are perpendicular in the usual Euclidean sense.

(8) In the Klein model two open chords are interpreted to be "perpendicular" if and only if they are perpendicular in the usual Euclidean sense.

(9) Inversion in a given circle maps all circles onto circles.

(10) Ultra-ideal points have no representation in the Poincaré models.

(11) Four points in the Euclidean plane from a harmonic tetrad if they are collinear and their cross-ratio equals one.

(12) If point O is outside circle δ and a tangent from O to δ touches δ at point T, then the power of O with respect to δ is equal to the square of the distance from O to T.

(13) Let point P lie on circle δ and let P′ and $\delta′$ be their inverses in another circle γ such that δ does not pass through the center of γ. Then the tangent to $\delta′$ at P′ is parallel to the tangent to δ at P.

(14) The inverse of the center of a circle δ is the center of the inverted circle $\delta′$.

(15) In order for the midpoint M of segment AB to have a harmonic conjugate with respect to AB, for all A and B, the Euclidean plane must be extended to the real projective plane by adding a line of points at infinity.

(16) If a statement in plane hyperbolic geometry holds when interpreted in the Klein or Poincaré model, then that statement is a theorem in hyperbolic geometry.

The following exercises (all of which are major exercises) will be divided into four categories: (1) K-exercises, on the Klein model; (2) P-exercises, on the Poincaré models and on circles; (3) H-exercises, on harmonic tetrads, theorems of Menelaus, Ceva, Gergonne, Desargues; (4) M-exercises, on motions of Euclidean and hyperbolic planes.

K-Exercises

K-1. Verify the interpretations of the incidence axioms, the betweenness axioms, and Dedekind's axiom for the Klein model (Archimedes' axiom follows from Dedekind's—Major Exercise 1, Chapter 3; see Exercise 31, Chapter 4, for the interpretation of B-4.)

K-2. (a) Let l be a diameter of γ and let m be an open chord of γ that does not meet l and whose end points differ from the end points of l. Draw a diagram showing the common perpendicular k to l and m in the Klein model. (Hint: use the pole of m and the description of perpendicularity in Case 1, p. 197.)

(b) Let l and m be intersecting open chords of γ. It is a valid theorem in hyperbolic geometry that for any two intersecting lines there exists a third line perpendicular to one of them and asymptotically parallel to the other (see Major Exercise 9, Chapter 6). Draw the two lines in the Klein model that are perpendicular to l and asymptotically parallel to m (on the left and right, respectively). This shows that the angle of parallelism can be any acute angle whatever. Explain.

(c) In the Euclidean plane any three parallel lines have a common transversal. Draw three parallel lines in the Klein model that do *not* have a common transversal.

K-3. (a) In the Klein model an ideal point and an ordinary point always determine a unique Klein line. Translate this back into a theorem in hyperbolic geometry about limiting parallel rays.

(b) Suppose the ultra-ideal points $P(l)$ and $P(m)$ are poles of Klein lines l and m, respectively. You saw in Figure 7.18 that the Euclidean line joining $P(l)$ and $P(m)$ need not cut through the circle γ, and hence need not determine a Klein line. Show that the only case in which there is a Klein line joining $P(l)$ and $P(m)$ is when l and m are divergently parallel.

(c) Suppose the ultra-ideal point $P(l)$ is the pole of a Klein line l and Ω is an ideal point; Ω is uniquely determined by a ray r in the direction of Ω. State the necessary and sufficient conditions on r and l in order that $P(l)$ and Ω determine a Klein line. Translate this into a theorem in hyperbolic geometry.

K-4. Given chords l and m of γ that are not diameters. Suppose the line extending m passes through the pole of l. Prove that the line extending l passes through the pole of *m*. (Hint: use either Equation 2, p. 215, or the theory of orthogonal circles.)

K-5. Use the Klein model to show that in the hyperbolic plane there exists a pentagon with five right angles and there exists a hexagon with six right angles. (Hint: begin with two lines having a common perpendicular. Locate the poles of these two lines, then draw an appropriate line through each of the poles, etc.) Does there exist, for any $n \geq 5$, an *n*-sided polygon with *n* right angles?

K-6. Justify the following construction of the Klein reflection A′ of A across *m*, which is simpler than the one in Figure 7.37: Let Λ be an end of *m* that is not collinear with A and the pole P of *m*. Join Λ to A and let this line cut *γ* again at Φ. Join Φ to P and let this line cut *γ* at Φ′. Then A′ is the intersection of $\overleftrightarrow{AP}$ with $\overleftrightarrow{\Lambda\Phi'}$. (See Figure 7.44.)

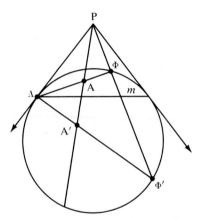

Figure 7.44

K-7. Given a segment AA′ in the Klein model. Show how to construct its hyperbolic midpoint with straightedge and compass (see Figures 7.37 and 7.38).

K-8. Construct triangles in the Klein model such that the perpendicular bisectors of the sides are (a) divergently parallel, and (b) asymptotically parallel. (See Exercise 13 and Major Exercise 7, Chapter 6.)

K-9. Prove the formula

$$F(z) = \frac{2z}{1 + |z|^2}$$

for the isomorphism F of the Poincaré model onto the Klein model (see Figure 7.31). What is the formula for the inverse isomorphism? Angle measure in the Klein model is defined so that F preserves angle measure; draw the diagram which illustrates this.

K-10. Let $A = (0,0)$, $B = (0,\frac{1}{2})$, and let l be the diameter of γ cut out by the x-axis.

(a) Find the Klein length $d'(AB)$.

(b) Find the coordinates of the point M on segment AB that represents its midpoint in the Klein model.

(c) Find the equation of the locus of points whose perpendicular Klein distance from l equals $d'(AB)$. (This locus is an "equidistant curve"—see Appendix A.)

K-11. Let Ω and Ω' be distinct ideal points and A an ordinary point. Let P be the pole of chord $\Omega\Omega'$, and let Euclidean ray $\overrightarrow{AP}$ cut γ at Σ. Prove that $A\Sigma$ represents the bisector of $\measuredangle\ \Omega A\Omega'$ in the Klein model (see Figure 7.45). Apply this result to justify the construction of the line of enclosure given in Major Exercise 8, Chapter 6. (Hint: use Theorem 6.6.)

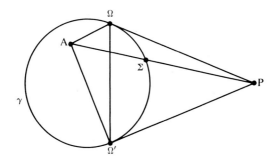

Figure 7.45

K-12. In Exercise 16, Chapter 6, you proved the theorem that the angle bisectors of a triangle in hyperbolic geometry (in fact, in neutral geometry) are concurrent. Using the construction of angle bisectors given in the previous exercise and the glossary of the Klein model, translate this theorem into a famous theorem in Euclidean geometry due to Brianchon (see Figure 7.46). This gives a hyperbolic proof of a Euclidean theorem (for a Euclidean proof, see Coxeter and Greitzer, p. 77).

K-13. It is a theorem in hyperbolic geometry that inside every trebly asymptotic triangle $\triangle\Sigma\Omega\Lambda$ there is a unique point G equidistant from all sides. Show that in the Klein model this theorem is a consequence of Gergonne's theorem in Euclidean geometry, which asserts that if the inscribed circle of $\triangle PQR$ touches the sides at points Λ, Σ, and Ω, then segments $P\Sigma$, $Q\Omega$, and $R\Lambda$ are concurrent (see Figure 7.47 and Exercise H-9). Show that $(\measuredangle\ \Lambda G\Sigma)° = 120°$ in the sense of degree measure for the Klein model. (Hint: to take care of the special case where one side of $\triangle\Sigma\Omega\Lambda$ is a diameter, apply a harmonic homology to transform to the case where Gergonne's theorem applies.)

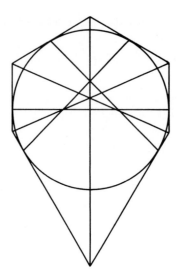

Figure 7.46

Brianchon's theorem.

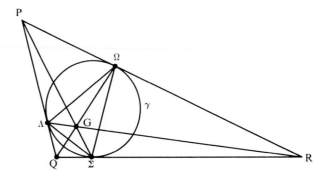

Figure 7.47

K-14. In order to express the Klein length $d'(AB) = \frac{1}{2}|\log (AB, PQ)|$ in terms of the coordinates (a_1, a_2) of A and (b_1, b_2) of B, prove that with a suitable ordering of the ends P and Q of the Klein line through A and B you have the formula

$$(AB, PQ) = \frac{a_1 b_1 + a_2 b_2 - 1 - \sqrt{(a_1 - b_1)^2 + (a_2 - b_2)^2 - (a_1 b_2 - a_2 b_1)^2}}{a_1 b_1 + a_2 b_2 - 1 + \sqrt{(a_1 - b_1)^2 + (a_2 - b_2)^2 - (a_1 b_2 - a_2 b_1)^2}}$$

(Hint: if A and B have complex coordinates z and w, then P and Q have complex coordinates $tz + (1 - t)w$ and $uz + (1 - u)w$, where t and u are roots of a quadratic equation $Dx^2 + 2Ex + F = 0$ expressing the fact that P and Q lie on the unit circle. Find the coefficients D, E and F and show that

$$(AB, PQ) = \frac{t(1-u)}{u(1-t)} = \frac{E + F - \sqrt{E^2 - DF}}{E + F + \sqrt{E^2 - DF}}.$$

K-15. Consider the statement in hyperbolic geometry, "Any three noncollinear points lie on the three respective sides of some trebly asymptotic triangle." Using the glossary for the Klein or Poincaré models, translate this into a statement in Euclidean geometry. Can you prove or disprove this statement? (See the remarks preceding Major Exercise 4, Chapter 6, for the definition of "trebly asymptotic triangle.")

P-Exercises

P-1. Using the glossary for the Poincaré disk model, translate the following theorems in hyperbolic geometry into theorems in Euclidean geometry:

(a) If two triangles are similar, then they are congruent.
(b) If two lines are divergently parallel, then they have a common perpendicular and the latter is unique.
(c) The fourth angle of a Lambert quadrilateral is acute.

P-2. Given an ordinary point A and an ideal point Ω in the Poincaré disk model. Show how to construct by straightedge and compass the P-line joining A to Ω. (Hint: use Proposition 7.5.)

P-3. Let δ be a circle with center C and α a circle not through C having center A. Let A′ be the inverse of A in δ and let circle α' be the image of α under inversion in δ. Prove that A′ is the inverse of C in α' and hence that A′ is not the center of α'. (Hint: show that any circle β through A′ and C is orthogonal to α' by observing that the image β' of β under inversion in δ is a line orthogonal to α.)

P-4. Let l be a Poincaré line that is not a diameter of γ; l is then an arc of a circle δ orthogonal to γ. Prove that hyperbolic reflection across l is represented in the Poincaré model by inversion in δ. (Hint: use Proposition 7.10 and the Corollary to Proposition 7.6.)

P-5. Let C be a point in the Poincaré disk model. Prove that a circle centered at C in the sense of hyperbolic geometry is represented in the Poincaré model by a Euclidean circle whose center is $\neq$C unless C coincides with the center O of γ. (Hint: first take C = O and use Lemma 7.4. Then map the set of circles centered at O onto the set of circles with C as hyperbolic center by reflection in the Poincaré line that is the Poincaré perpendicular bisector of the Poincaré segment OC. Apply Exercises P-3 and P-4.)

P-6. In the hyperbolic plane with some given unit of length, the distance d for which the angle of parallelism $\Pi(d)^\circ = 45^\circ$ is called *Schweikart's constant*. Schweikart was the first to notice that if $\triangle ABC$ is an isosceles right triangle with base BC, then the length of the altitude from A to BC is bounded by this constant, which is the least upper bound of the lengths of all such altitudes. Prove that for the length function we have defined for the Poincaré disk model, Schweikart's constant equals $\log(1 + \sqrt{2})$ (see Figure 7.48). (Hint: Schweikart's constant is the Poincaré length d of segment OP in Figure 7.48. Show that the Euclidean length of OP is $\sqrt{2} - 1$ and apply Lemma 7.4 to solve for d.)

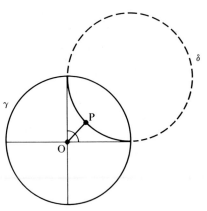

Figure 7.48

P-7. Let α be a circle with center A and radius of length r and β a circle with center B and radius of length s. Assume $A \neq B$ and let C be the unique point on $\overleftrightarrow{AB}$ such that $\overline{AC}^2 - \overline{BC}^2 = r^2 - s^2$. The line through C perpendicular to $\overleftrightarrow{AB}$ is called the *radical axis* of the two circles.

(a) Prove (e.g., by introducing coordinates) that C exists and is unique, and that for any point P different from A and B, P lies on the radical axis if and only if $\overline{PA}^2 - \overline{PB}^2 = r^2 - s^2$.

(b) For any point X outside both α and β, let T be a point of α such that $\overleftrightarrow{XT}$ is tangent to α at T; similarly let U on β be a point of tangency for $\overleftrightarrow{XU}$. Prove that $\overline{XT} = \overline{XU}$ if and only if X lies on the radical axis of α and β.

(c) Prove that if α and β intersect in two points P and Q, $\overleftrightarrow{PQ}$ is their radical axis.

(d) Prove that if α and β are tangent at point C, the radical axis is the common tangent line through C. (Hint: use the Pythagorean theorem and some algebra.)

(e) Let X be a point outside both α and β. Prove that X lies on the radical axis of α and β if and only if X has the same power with respect to α and β (see Lemma 7.1).

P-8. Given two nonintersecting, nonconcentric circles α and β with centers A and B, respectively. Justify the following straightedge-and-compass construction of the radical axis of α and β: Draw any circle δ that cuts α in two points A' and A" and cuts β in two points B' and B". Then $\overleftrightarrow{A'A''}$ and $\overleftrightarrow{B'B''}$ intersect in a point P that lies on the radical axis; the latter is therefore the perpendicular to $\overleftrightarrow{AB}$ through P. (Hint: draw tangents PS, PT, and PU from P to δ, α, and β and apply Exercise P-7 (b) and (c) to show that $\overline{PT} = \overline{PS} = \overline{PU}$. See Figure 7.49.)

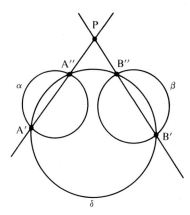

Figure 7.49

P-9. Use Exercise P-7 to verify by a straightedge-and-compass construction that in the Poincaré model two divergently parallel Poincaré lines have a common perpendicular. (Hint: there are four cases to consider, depending on whether the Poincaré line is a diameter of γ or an arc of a circle α orthogonal to γ, depending on whether radical axes intersect or not, etc. One case is illustrated in Figure 7.50.) In case the radical axes are parallel, use the fact that the perpendicular bisector of a chord of a circle passes through the center of the circle (Exercise 17, Chapter 4).

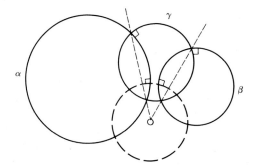

Figure 7.50

P-10. Given any Poincaré line l and any Poincaré point P not on l. Construct the two rays from P in the Poincaré model that are limiting parallel to l. (If l is an arc of a circle α orthogonal to γ and intersecting γ at A_1 and A_2, then the problem amounts to constructing a circle β_i through P that is orthogonal to γ and tangent to α at A_i for each of $i = 1, 2$. Use Proposition 7.5.)

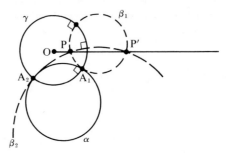

Figure 7.51

P-11. Given an acute angle in the Poincaré model. Construct the unique Poincaré line that is perpendicular to a given side of this angle and limiting parallel to the other. This shows that the angle of parallelism can be any acute angle whatever. (Hint: if both Poincaré lines are arcs of orthogonal circles α and β, let P′ be the intersection with γ of the part of α containing the given ray, and let P be the other intersection with γ of $\overleftrightarrow{P'B}$, B being the center of β. Show that P and P′ are inverse points in circle β, then find the point of intersection of the tangents to γ at P and P′. Compare with Major Exercise 9, Chapter 6.)

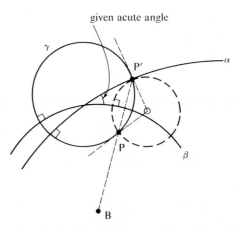

given acute angle

Figure 7.52

P-12. Given circle γ with center O. For any point P $\neq$ O, if P′ is the inverse of P in γ, then the line through P′ that is perpendicular to $\overleftrightarrow{OP}$ is called the *polar* of P with respect to γ and will be denoted $p(P)$. When P lies outside γ, its polar joins the points of contact of the two tangents to γ from P (see Figure 7.21). When P lies on γ, its polar is the tangent to γ at P, and this is the only case in which P lies on $p(P)$. Prove the following duality property: B lies on $p(A)$ if and only if A lies on $p(B)$. (Hint: if B lies on $p(A)$, let B′ be the foot of the perpendicular from A to $\overleftrightarrow{OB}$. Show that $\triangle OAB'$ is similar to $\triangle OBA'$ and deduce that B′ is the inverse of B in γ. For the significance of this operation of polar reciprocation for the theory of conics, see Coxeter and Greitzer, Chapter 6.)

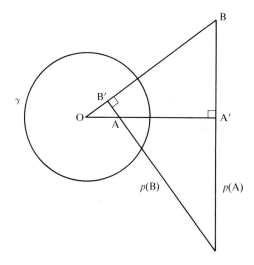

Figure 7.53

P-13. We define three types of *coaxal pencils of circles* as follows:

(1) Given a line t and a point C on t. The corresponding *tangent coaxal pencil* consists of all circles tangent to t at C.

(2) Given two points A and B. The corresponding *intersecting coaxal pencil* consists of all the circles which pass through both A and B, and A and B are the *limiting points* of this pencil.

(3) Given a circle γ and a line t not meeting γ. The corresponding *nonintersecting coaxal pencil* consists of γ and all other circles δ such that t is the radical axis of γ and δ.

Prove the following:

(a) Any two nonconcentric circles belong to a unique coaxal pencil.

(b) Given a coaxal pencil C. All pairs of circles belonging to C have the same radical axis, and the centers of all circles in C lie on a line perpendicular to this radical axis called the *line of centers* of C.

(Hint: See Exercise P-7.)

P-14. Prove the following:

(a) The set of all circles orthogonal to two given tangent circles γ and δ is the tangent coaxal pencil whose line of centers is the common tangent t to γ and δ.

(b) The set of all circles orthogonal to two given nonintersecting circles γ and δ is the intersecting coaxal pencil whose line of centers is the radical axis t of γ and δ and whose limiting points are the two points at which every member of this pencil cuts the line joining the centers of γ and δ.

(c) The set of all circles orthogonal to two given circles γ and δ intersecting at A and B is the nonintersecting coaxal pencil whose line of centers is $\overleftrightarrow{AB}$ and whose radical axis is the perpendicular bisector of AB. (See Figure 7.54.)

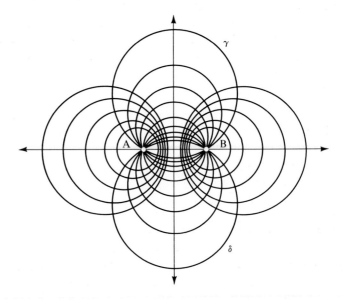

Figure 7.54

P-15. Given three circles α, β, and γ. Is there always a fourth circle δ orthogonal to all three of them? If so, is δ unique? (Hint: consider the radical axes of the three pairs of circles obtained from the three given circles; the center of δ must lie on all three radical axes and must lie outside the three circles.)

P-16. Given a circle γ with center O.

(a) Given $P \neq O$ and P' its inverse in γ. Prove that inversion in γ maps the pencil of lines through P' onto the intersecting coaxal pencil of circles through O and P and maps the orthogonal pencil of concentric circles centered at P' onto the nonintersecting coaxal pencil of circles whose radical axis is the perpendicular bisector of OP.

(b) Given a line l through O. Prove that inversion in γ maps the pencil of lines parallel to l onto the pencil of circles tangent to l at O.

P-17. *The inversive plane* is obtained from the Euclidean plane by adjoining a single point at infinity ∞, which by convention lies on every Euclidean line but does not lie on any Euclidean circle. By a "circle" we mean either an ordinary Euclidean circle or a line in the inversive plane. Two parallel Euclidean lines meet at ∞ when extended to inversive lines; as "circles" they will be considered to be tangent at ∞. Given an ordinary circle γ with center O, define the inverse of O in γ to be ∞. By inversion in a "circle" we mean either inversion in an ordinary circle or reflection across a line. Prove the following:

(a) Inversion in a given "circle" maps "circles" onto "circles."
(b) If A and B are inverse to each other in a "circle" α, and if under inversion in another "circle" β they map to A′, B′, α′, then A′ and B′ are inverse to each other in α′. (Hint for b: show that any "circle" γ′ through A′ and B′ is orthogonal to α′ by observing that inversion preserves orthogonality— use Propositions 7.5 and 7.9.)

P-18. In addition to the tangent, intersecting, and nonintersecting coaxal pencils of circles defined in Exercise P-13, define three further pencils of "circles" in the inversive plane as follows:

(4) All the circles having a given point as center.
(5) All the lines passing through a given ordinary point.
(6) A given line and all lines parallel to it.

Furthermore, given a coaxal pencil of circles, we will consider its radical axis as one more "circle" belonging to the pencil. Prove the following:

(a) Two distinct "circles" belong to a unique pencil of "circles."
(b) A pencil of "circles" is invariant as a set under inversion in any "circle" in the pencil. (Hint for b: the statement is obvious for the three new types of pencils just introduced. For the three coaxal types, use the two preceding exercises.)

P-19. Show that an isomorphism mapping the Poincaré upper half-plane model onto the Poincaré disk model is given by $\Phi(z) = (i - z)/(i + z)$, where $i^2 = -1$ and z is a complex variable.

Prove that every reflection in the upper half-plane model can be expressed in the form $z \to (a\bar{z} + b)/(c\bar{z} - a)$, where $a, b,$ and c are real numbers satisfying $a^2 + bc = 1$ and $\bar{z}$ denotes the complex conjugate of z; conversely, each such transformation is a reflection across a suitable Poincaré line. Prove that a product of two reflections in this model has the form $z \to (az + b)/(cz + d)$, where $a, b, c,$ and d are real numbers satisfying $ad - bc = 1$; conversely, each such transformation is a product of two reflections. (Hint: see Artzy, p. 54. This exercise shows that the group of direct motions of the hyperbolic plane is isomorphic to PSL$(2, \mathbf{R})$; see Exercise M-23.)

H-Exercises

H-1. Let M be the midpoint of AB, $r = \overline{MA}$, and let C, D on $\overleftrightarrow{AB}$ lie on the same side of M, with A, B, C, D distinct. Then C and D are harmonic conjugates with respect to AB if and only if $r^2 = (\overline{MD})(\overline{MC})$.

H-2. If γ and δ are orthogonal circles, AB is a diameter of γ, and δ cuts $\overleftrightarrow{AB}$ in points C and D, then C and D are harmonic conjugates with respect to AB; conversely, if a diameter of one circle is cut harmonically by a second circle, then the two circles are orthogonal (see Figure 7.55). (Hint: if T is a point of intersection of γ and δ, use Lemma 7.1 to show that the circles are orthogonal if and only if $(\overline{OT})^2 = (\overline{OC})(\overline{OD})$. Now apply Exercise H-1.)

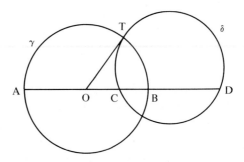

Figure 7.55

H-3. Given three collinear points A, B, and C. Prove that the fourth harmonic point D is the inverse of C in the circle having AB as diameter. (Hint: use Exercise H-2 and Proposition 7.5.)

H-4. *Sensed magnitudes.* Given two points A, B. Assign arbitrarily an order (i.e., a direction) to $\overleftrightarrow{AB}$ (Exercise 7, Chapter 3). Then the length of AB will be considered positive or negative according to whether the direction from A to B is the positive or negative direction on the line. We will denote this signed length by $\underline{AB}$, so that we have $\underline{AB} = -\underline{BA}$. If C is a third point on directed line $\overleftrightarrow{AB}$, we define the signed ratio in which C divides AB to be $\underline{AC}/\underline{CB}$.

(a) Prove that this signed ratio is independent of the direction assigned to the line and that point C is uniquely determined by this ratio. (Note that C would not be uniquely determined by the unsigned ratio.)

(b) Given parallel lines *l* and *m*. Let transversals *t* and *t'* cut *l* and *m* in B, C and B', C', respectively, and let *t* meet *t'* at point A. Prove that AB/BC = AB'/B'C' (see Exercise 18, Chapter 5).

H-5. *Theorem of Menelaus.* Given △ABC and points D on B⃡C, E on C⃡A, and F on A⃡B that do not coincide with any of the vertices of the triangle. Define the *linearity number* by [ABC/DEF] = (AF/FB)(BD/DC)(CE/EA). Then a necessary and sufficient condition for D, E, and F to be collinear is that [ABC/DEF] = −1. (Hint: if D, E, and F lie on a line *l*, let the parallel *m* to *l* through A cut B⃡C at G. Use Exercise H-4 to get CE/EA = CD/DG and AF/FB = GD/DB. and deduce that the linearity number is −1. Conversely, use Exercise H-4 to show that E⃡F cannot be parallel to B⃡C. If these lines meet at D', use the first part of the proof and the hypothesis to show that BD/DC = BD'/D'C and apply Exercise H-4a.

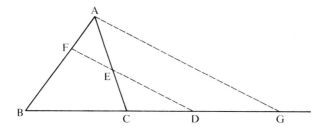

Figure 7.56

H-6. *Theorem of Ceva.* Given △ABC and a third point D (respectively E, F) on B⃡C (resp. on A⃡C, A⃡B). Then the three lines A⃡D, B⃡E, and C⃡F are either concurrent or parallel if and only if [ABC/DEF] = +1 (see Figure 7.57). (Hint: suppose that the three lines meet at O; apply Menelaus' theorem to △ADB and △ADC to obtain two different expressions for OD/AO, then

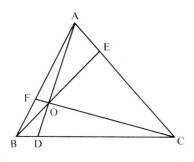
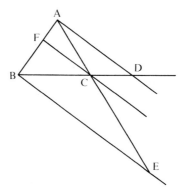

Figure 7.57

divide one expression by the other to see that the linearity number is $+1$. If the three lines are parallel, apply Exercise H-4b. Conversely, if the linearity number is $+1$ and the three lines are not parallel, let $\overleftrightarrow{BE}$ and $\overleftrightarrow{CF}$, e.g., meet at O, and let $\overleftrightarrow{AO}$ meet $\overleftrightarrow{BC}$ at D′. Use the first part of the proof and the hypothesis to show that $\underline{BD}/\underline{DC} = \underline{BD'}/\underline{D'C}$ and apply Exercise H-4a.)

H-7. Given four collinear points A, B, C, and D. Define their *signed cross-ratio* (AB, CD) by $(\underline{AB}, \underline{CD}) = (\underline{AC}/\underline{CB})/(\underline{AD}/\underline{DB})$.
  (a) Prove that ABCD is a harmonic tetrad if and only if $(\underline{AB}, \underline{CD}) = -1$.
  (b) Prove that signed cross-ratios are preserved by perspectivities and parallel projections (see Lemma 7.5 and the parallel projection theorem, p. 137).

H-8. Prove that ABCD is a harmonic tetrad if and only if $1/\underline{AB} = \frac{1}{2}(1/\underline{AC} + 1/\underline{AD})$.

H-9. Suppose the inscribed circle of $\triangle ABC$ touches sides BC, CA, and AB at D, E, and F, respectively. Prove that AD, BE, and CF are concurrent in a point G called the *Gergonne point* of $\triangle ABC$. (Hint: by Exercise 16, Chapter 6, the center I of the inscribed circle lies on all three angle bisectors; this gives three pairs of congruent right triangles that can be used to verify the criterion of Ceva's theorem.)

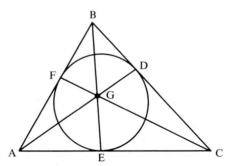

Figure 7.58

H-10. Use the theorem of Menelaus to prove Desargues' theorem as stated in Major Exercise 11, Chapter 2. (Hint: referring to Figure 2.10, p. 56, apply Menelaus' theorem to $\triangle BCP$, $\triangle CAP$, and $\triangle ABP$, then multiply the three equations to get $[ABC/RST] = -1$. Now apply Menelaus' theorem once more.)

H-11. The theorems of Menelaus and Ceva can be applied to prove famous theorems of Pappus and Pascal and to prove the existence of special points of a triangle. Report on these results, using Kay or Coxeter and Greitzer as references.

M-Exercises

The following are exercises in neutral geometry unless otherwise indicated.

M-1. We define a *motion* or *isometry* of the plane to be a one-to-one transformation T of the plane onto itself that preserves lengths: For any two points A and B, if $A' = T(A)$ and $B' = T(B)$, then $AB = \overline{A'B'}$. Prove that the motions form a *group*, i.e., that the following three conditions hold:

(a) If S and T are motions, then so is the composite ST.

(b) If T is a motion then so is its inverse T^{-1}.

(c) The *identity* transformation I defined by $I(A) = A$ for all points A is a motion.

M-2. Given a line m. Recall the definition of the reflection R_m across m given on p. 218. Prove that R_m is a motion and that it is equal to its own inverse.

M-3. Prove that motions preserve collinearity, betweenness, and congruence of both segments and angles (use Theorem 4.3).

M-4. T is a motion and l and m are lines. Since T preserves collinearity, the images Tl and Tm of l and m under T are also lines. Prove the following:

(a) l and m meet at $P \Leftrightarrow Tl$ and Tm meet at TP.

(b) l and m are parallel $\Leftrightarrow Tl$ and Tm are parallel.

(c) l and m have a common perpendicular $t \Leftrightarrow Tl$ and Tm have a common perpendicular Tt.

(d) In the hyperbolic plane, l and m are asymptotically parallel $\Leftrightarrow Tl$ and Tm are asymptomatically parallel (use Theorem 6.7).

(e) T maps each half-plane determined by l onto a half-plane determined by Tl.

M-5. For the Euclidean plane we defined a *dilation* with center O and ratio k on p. 206. Prove that this transformation preserves collinearity, betweenness, and congruence but is not a motion unless $k = 1$ and it is the identity. (Hint: introduce a rectangular coordinate system and use the Pythagorean theorem.) Prove that the set of all motions, dilations, and their composites is a group of transformations of the Euclidean plane called the group of *similitudes*. Prove that if S is a similitude, then every triangle is similar to its image under S. Prove that in the hyperbolic plane a transformation with this last property must be a motion (use Theorem 6.2).

M-6. Given two points A and B on line l. Suppose motions S and T agree at A and B, i.e., $SA = TA$ and $SB = TB$. Prove that for every point P on l, $SP = TP$. In particular, deduce that if $TA = A$ and $TB = B$, then T leaves every point on l fixed.

M-7. Given congruent triangles $\triangle ABC \cong \triangle A'B'C'$. Prove that there is a unique motion T such that $TA = A'$, $TB = B'$, and $TC = C'$. (Hint: use Exercise M-6 and Major Exercise 2, Chapter 4, to construct T. There are four cases to consider in proving that T preserves length.)

M-8. Let T be a motion that is not the identity and that fixes two points on line l. Prove that $T = R_l$ (use the uniqueness part of Exercise M-7).

M-9. If two lines l and m intersect at a point A, then the composite $R_l R_m$ of the reflections across l and m is called a *rotation* about A; in the special case where $l \perp m$ it is called the *half-turn* H_A about A. Prove that A is the only fixed point of this rotation. (Hint: if B were also fixed, let $B' = R_l B = R_m B$; show that $\overleftrightarrow{BB'}$ is perpendicular to both l and m.) Conversely, prove that if T is a motion that has exactly one fixed point A, then T is a rotation about A. (Hint: choose any other points B and C not collinear with A, and let $B' = TB$, $C' = TC$, $m = \overleftrightarrow{AB}$, and let l be the perpendicular bisector of BB'. Prove that $T = R_l R_m$ by showing that $R_l T$ fixes A and B.)

M-10. Prove that a motion T that has no fixed points is a product of two or three reflections. Deduce from this and the previous two exercises that a motion that is not the identity is either a reflection or the product of two or three reflections. (Hint: for any point A, show that if $A' = TA$ and l is the perpendicular bisector of AA', then $R_l T$ leaves A fixed and hence is either a rotation about A or a reflection in a line through A.)

M-11. Given lines l and m intersecting at point A. Prove that the rotations $R_l R_m$ and $R_m R_l$ are inverse to each other. Prove that $R_l R_m = R_m R_l \Leftrightarrow l \perp m$. In the case l and m are not perpendicular, prove that if B' and C' are the images under $R_l R_m$ of any two points B and C different from A, then $(\measuredangle CAC')° = (\measuredangle BAB')°$.

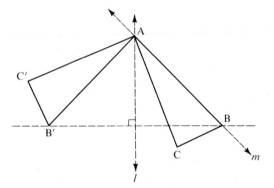

Figure 7.59

M-12. Let R be a rotation about a point A and let l be any line through A. Prove that there is another unique line k through A such that $R = R_k R_l$. (Hint: review your solution to Exercise M-9.) Deduce that if S is another rotation about A, then the composite RS is also a rotation about A. (Hint: let $S = R_l R_m$. Write $R = R_k R_l$ from the first part and multiply.) Similarly, show that $RS = SR$.

M-13. For a line t, a motion T is called a *translation along* t if there are two perpendiculars l and m to t such that $T = R_l R_m$. Suppose that l cuts t at A and m cuts t at B. Prove that for any point Q on t, if $Q' = TQ$, then $\overline{QQ'} = 2(\overline{AB})$. Let P be any point not on t, Q the foot of the perpendicular from P to t, and $P' = TP$. Prove that P' is on the same side of t as P, and that $\overline{PP'} = \overline{QQ'}$ or $\overline{PP'} > \overline{QQ'}$, depending on whether the geometry is Euclidean or hyperbolic (see Figure 7.60). (Hint: show that $\square PQQ'P'$ is a Saccheri quadrilateral with QQ' as base, and use Exercise 1, Chapter 6.) Deduce that a translation has no fixed points. Prove also that $T = H_B H_A$ (see Exercise M-9).

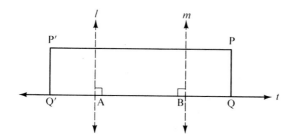

Figure 7.60

M-14. Let T be a translation along a line t and let k be any perpendicular to t. Prove that there is another, unique perpendicular h to t such that $T = R_h R_k$. Deduce that a composite ST of two translations along t is a translation along t and $ST = TS$ (as in Exercise M-12). Prove also that T^{-1} is a translation along t. (Hint: if k cuts t at K and $K' = TK$, show that the perpendicular bisector h of KK' is the desired line by observing that $R_h T$ fixes every point of k.)

M-15. For a motion T, a line l is *invariant* under T if $Tl = l$. (This does not necessarily mean that T leaves every point of l fixed.) Prove the following:

(a) The invariant lines of a reflection R_m are its axis m and all the lines perpendicular to m.
(b) The invariant lines of a half-turn H_A are all the lines through A.
(c) A rotation that is not a half-turn has no invariant lines.

(d) In the Euclidean plane the invariant lines of a translation T along t are t and all the lines u parallel to t, and T is also a translation along u.

(e) In the hyperbolic plane, if T is a translation along t, then t is the unique invariant line of T. (Hint: assume on the contrary that T had another invariant line u. Dispose successively of the possibilities u meets t, u is divergently parallel to t, and u is asymptotically parallel to t, using the absence of fixed points and Theorems 6.4 and 6.5.)

In Exercises M-16 through M-18 the geometry is assumed to be hyperbolic.

M-16. For a motion T, if asymptotically parallel lines l and m "meet" in an ideal point Σ, then Tl is asymptotically parallel to Tm, and if Σ' is the ideal point at which Tl "meets" Tm, we set $T\Sigma = \Sigma'$. If $\Sigma' = \Sigma$, we say Σ is an *ideal fixed point* of T. Prove the following:

(a) If Ω and Σ are the ideal ends of a line m, then Ω and Σ are both fixed by the reflection R_m across m.

(b) If a motion fixes an ideal point Σ and an ordinary point A, then it must be the reflection in the line "joining" A to Σ; consequently a rotation has no ideal fixed points. (Hint: show that a rotation about A leaves no ray emanating from A invariant.)

(c) A translation along a line t leaves an ideal point Σ fixed if and only if Σ is an end of t. (Hint: Σ is fixed if and only if the line joining Σ to either end of t is invariant; use Exercise M-15.)

M-17. If l is asymptotically parallel to m, then the composite $P = R_l R_m$ is called a *parallel displacement* about the ideal point Σ at which l "meets" m. Prove the following:

(a) P has no ordinary fixed points (assume on the contrary that P fixes A and show that if $A' = R_l A = R_m A$, then $\overleftrightarrow{AA'}$ is perpendicular to both l and m).

(b) For any point A, let $A'' = PA$ and let k be the line joining A to Σ. Prove that Σ "lies on" the perpendicular bisector h of AA'' and that $P = R_h R_k$ (use Major Exercise 7, Chapter 6, to show that $R_h P$ fixes A and Σ—see Figure 7.61).

(c) P has no invariant lines (if t is invariant, choose A on t and, with h and k as in part b of the exercise, show that $h \perp t$ so that $R_h t = t$, whence $R_k t = Pt = t$ and t is perpendicular to both h and k).

(d) The only ideal fixed point of P is Σ (use part c of the exercise).

(e) The set of all parallel displacements about Σ together with the identity is a group of transformations, and this group is commutative (i.e., $PS = SP$).

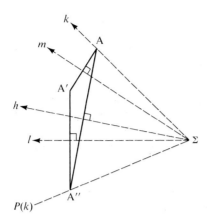

Figure 7.61

M-18. Let k and k' be asymptotically parallel lines "meeting" at ideal point Σ. Prove that there is a unique parallel displacement P about Σ such that $Pk = k'$. (Hint: let Ω and Ω' be the other ends of k and k', let $t = \Omega\Omega'$, and let h be the unique perpendicular to t through Σ (Exercise K-3c). Show that $P = R_h R_k$ is the solution. For uniqueness use Exercise M-17c.)

We return to neutral geometry in the remaining exercises.

M-19. Let T and T' be translations along distinct lines t and t'. Prove the following:

(a) If t meets t' in a point A, then TT' is a translation along a line t'' that intersects both t and t'. (Hint: use Exercises M-13 and M-14 to write $T = H_B H_A$ and $T' = H_A H_C$, and take $t'' = \overleftrightarrow{BC}$.) Prove that if the geometry is Euclidean, then $TT' = T'T$, whereas if the geometry is hyperbolic, $TT' \neq T'T$. (Hint: show that $T'T$ is a translation along a different line in hyperbolic geometry and apply Exercise M-15e. For the Euclidean case, use Exercise M-15d to get $T' = H_B H_D$, as in Figure 7.62.)

(b) If the geometry is Euclidean and t is parallel to t', then TT' is a translation along t and $TT' = T'T$.

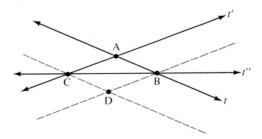

Figure 7.62

(c) If the geometry is hyperbolic and t and t' have a common perpendicular m, then $TT' = R_l R_n$, where $l \perp t$ and $n \perp t'$, $l \neq n$. (Hint: use Exercise M-14 to write $T = R_l R_m$ and $T' = R_m R_n$.) Deduce that if l and n do not have a common perpendicular, then TT' is not a translation. By choosing all these lines appropriately, show that TT' could be a rotation or a parallel displacement.

(d) If the geometry is hyperbolic and t is asymptotically parallel to t', then TT' cannot be a rotation. (Hint: use Exercise M-16b.)

M-20. A *glide* G along a line t is a product $R_t T$ of a translation T along t and the reflection across t. Prove the following:

(a) $R_t T = T R_t$.

(b) $H_A R_m = G = R_l H_B$, where l and m are perpendiculars to t at A and B such that $T = R_l R_m$, and H_A and H_B are half-turns.

(c) Conversely, a product $R_l H_B$ of a reflection and a half-turn is either a glide along the perpendicular t to l through B if B does not lie on l, or is the reflection in t if B does lie on l.

(d) The glide G maps each side of t onto the opposite side.

(e) Glide G has no fixed points.

(f) The only invariant line of G is t.

M-21. We say that three lines l, m, and n *belong to a pencil* if one of the following holds:

(1) They are concurrent.

(2) They have a common perpendicular.

(3) They are asymptotically parallel in the direction of the same ideal point (hyperbolic plane only).

Prove the following theorem: Suppose that motion S is a product of three reflections, $S = R_l R_m R_n$. If l, m, and n belong to a pencil, then S is a reflection in a line of that pencil; otherwise, S is a glide. (Hint: for the three types of pencils, use Exercises M-12, M-14, and M-17b, respectively. If the lines do not belong to a pencil, choose any point A on l. Following the terminology of pp. 198–199 ("Perpendicularity in the Beltrami-Klein Model"), let m' be the line "joining" A to the ordinary, ideal, or ultra-ideal point at which m "meets" n. Then m', m, and n belong to a pencil, and $R_{m'} R_m R_n = R_{n'}$. Let B be the foot of the perpendicular k from A to n', and let h be the line such that $R_l R_{m'} R_k = R_h$. Show that B does not lie on h and that $S = R_h H_B$. See Figure 7.63.)

M-22. Prove that every product $R_l R_m H_A$ can be rewritten in the form $R_h R_k$. (Hint: let n be a line through A in the pencil determined by l and m; use Exercise M-21 to write $R_l R_m R_n = R_h$ and take k to be the perpendicular to n through A.)

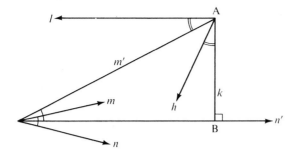

Figure 7.63

M-23. A motion is called *direct* or *orientation-preserving* if it is a product of two reflections or else is the identity. A motion is called *opposite* or *orientation-reversing* if it is a reflection or a glide. Prove the following theorem:

Every motion is either direct or opposite and not both. The product of two opposite motions is direct. The set of direct motions is a group of transformations. The product of a direct and an opposite motion is opposite.

(Hint: use practically all the preceding exercises, especially M-22 and M-21, to reduce products of more than three reflections to products of three or less.)

M-24. Prove that for congruent segments AB $\cong$ A'B' there is a unique direct motion S such that SA = A' and SB = B'. (Hint: use Exercises M-7 and M-23.)

M-25. Two plane figures are called *congruent* if there is a motion mapping one figure onto the other. Prove that all trebly asymptotic triangles in the hyperbolic plane are congruent to one another. (Hint: use Exercise K-13 and the fact that there is a motion mapping one 120° angle onto another.)

M-26. If A is an ordinary point and Σ is an ideal point in the hyperbolic plane, the set of all points PA as P runs through all the parallel displacements about Σ is called the *horocycle* through A centered at Σ. Prove that all horocycles are congruent to one another. (Hint: first prove that any two rays AΣ and BΩ are congruent to each other.)

8 | Philosophical Implications

Mathematics is a yoga and yoga becomes mathematics; there is no end to its depths.
Baba Hari Dass

What is the Geometry of Physical Space?

We have shown that if Euclidean geometry is consistent, so is hyperbolic geometry, since we can construct models for it within Euclidean geometry. Conversely, it can be proved that if hyperbolic geometry is consistent, so is Euclidean geometry, for the "horocycles" on the "horosphere" in hyperbolic space form a model of the lines on the Euclidean plane (see Kulczycki, §7). *Thus, the two geometries are equally consistent.*

You may grant now that, logically speaking, hyperbolic geometry deserves to be put on an equal footing with Euclidean geometry. But you may also feel that hyperbolic geometry is just an amusing intellectual pastime, whereas Euclidean geometry accurately represents the physical world we live in and is therefore far more important. Let's examine this idea a little more closely.

Certainly, engineering and architecture are evidence that Euclidean geometry is extremely useful for ordinary measurement of distances that are not too large. However, the representational accuracy of Euclidean geometry is less certain when we deal with larger distances. For example, let us interpret a "line" physically as the path traveled by a light ray. We could then consider three widely separated light sources forming a physical triangle. We would want to measure the angles of this physical triangle in order to verify whether the sum is 180° or not (such an experiment would presumably settle the question of whether space is Euclidean or hyperbolic).

Gauss allegedly performed this experiment, using three mountain peaks as the vertices of his triangle. The results were inconclusive. Why? Because any physical experiment involves experimental error. Our instruments are never completely accurate. Suppose the sum did turn out to be 180°. If the error in our measurement were at most 1/100 of a degree, we could conclude only that the sum was between 179.99° and 180.01°. We could never be sure that it actually was 180°.

Suppose, on the other hand, that measurement gave us a sum of 179°. Although we could conclude only that the sum was between 178.99° and 179.01°, we would be certain that the sum was less than 180°. In other words, the only conclusive result of such an experiment would be that space is hyperbolic! The inconclusiveness of Gauss' experiment shows only that if space is hyperbolic, the defects of terrestrial triangles are extremely small.* The experiment can then be shifted to astronomical triangles formed by stars. Even there the measurements have been inconclusive.

To repeat the point: because of experimental error, a physical experiment can never prove conclusively that space is Euclidean—it can prove only that space is non-Euclidean.

The discussion can be made more subtle. We must question the nature of our instruments—aren't they designed on the basis of Euclidean assumptions? We must question our interpretation of "lines"—couldn't light rays travel on curved paths? We must question whether space, especially space of cosmic dimensions, cannot be described by geometries other than these two.

The latter is in fact our present scientific attitude. According to Einstein, space and time are inseparable and the geometry of space-time is affected by matter, so that light rays are indeed curved by the gravitational attraction of masses. Space is no longer conceived of as an empty Newtonian box whose contours are unaffected by the rocks put into it. The problem is much more complicated than Euclid or Lobachevsky ever imagined—neither of their geometries is adequate for our present conception of space. This does not diminish the historical importance of non-Euclidean geometry. Einstein said, "To this interpretation of geometry I attach great importance, for should I not have been acquainted with it, I never would have been able to develop the theory of relativity."†

---

* If the measurement gave us a sum of 181° with error at most .01°, we would conclude that space is elliptic.

† See George Gamow's article, "The Evolutionary Universe," *Scientific American,* September, 1956, which tells how Einstein developed a geometry appropriate to general relativity from the ideas of Georg Friedrich Bernhard Riemann (1826–1866).

Here is the famous response of Poincaré to the question of which geometry is true:*

> If geometry were an experimental science, it would not be an exact science. It would be subjected to continual revision . . . *The geometrical axioms are therefore neither synthetic a priori intuitions nor experimental facts. They are conventions.* Our choice among all possible conventions is guided by experimental facts; but it remains free, and is only limited by the necessity of avoiding every contradiction, and thus it is that postulates may remain rigorously true even when the experimental laws which have determined their adoption are only approximate. In other words, *the axioms of geometry* (I do not speak of those of arithmetic) *are only definitions in disguise.* What then are we to think of the question: Is Euclidean Geometry true? It has no meaning. We might as well ask if the metric system is true and if the old weights and measures are false; if Cartesian coordinates are true and polar coordinates false. *One geometry cannot be more true than another: it can only be more convenient.* [italic added]

You may think that Euclidean geometry is the most convenient—it is for ordinary engineering, but not for the theory of relativity. Moreover, R. K. Luneburg contends that visual space, the space mapped on our brains through our eyes, is most conveniently described by hyperbolic geometry.†

Philosophers are still arguing about Poincaré's philosophy of conventionalism. One school, which includes Newton, Helmholtz, Russell, and Whitehead, contends that space has an intrinsic metric or standard of measurement. The other school, which includes Riemann, Poincaré, Clifford, and Einstein, contends that a metric is stipulated by convention. The discussion can become very subtle (see Grünbaum).

What is Mathematics About?

The preceding discussion sheds new light on what geometry, and in general, mathematics, is about. Geometry is not about light rays, but the path of a light ray is one possible *physical interpretation* of the undefined

---

* *Science and Hypothesis,* New York: Dover, 1952, p. 50; originally published in French, 1902.

† See his book *Mathematical Analysis of Binocular Vision,* and his article in the *Optical Society of America Journal,* October, 1950, p. 629. See also the articles by O. Blank in that same journal, December, 1958, p. 911, and March, 1961, p. 335.

Albert Einstein

geometric term "line." Bertrand Russell once said that "mathematics is the subject in which we do not know what we are talking about nor whether what we say is true." This is because certain primitive terms, such as "point," "line," and "plane," are undefined and could just as well be replaced with other terms without affecting the validity of results. Instead of saying "two points determine a unique line," we could just as well write "two alphas determine a unique beta." Despite this change in terms, the proofs of all our theorems would still be valid, because correct proofs do not depend on diagrams; they depend only on stated axioms and the rules of logic. Thus, geometry is a purely *formal* exercise in deducing certain conclusions from certain formal premises. Mathematics makes statements of the form "if . . . then"; it does not say anything about the meaning or truthfulness of the hypotheses. The primitive notions (such as "point" and "line") appearing in the hypotheses are implicitly defined by these axioms, by the rules as it were that tell us how to play the game.*

The formalist viewpoint just stated is a radical departure from the older notion that mathematics asserts "absolute truths," a notion that was destroyed once and for all by the discovery of non-Euclidean geometry. This discovery has had a liberating effect on mathematicians, who now feel free to invent any set of axioms they wish and deduce conclusions from them. In fact, this freedom may account for the great increase in the scope and generality of modern mathematics. In a 1961 address,† Jean Dieudonné remarked on Gauss' discovery of non-Euclidean geometry:

[It] was a turning point of capital significance in the history of mathematics, marking the first step in a new conception of the relation between the real world and the mathematical notions supposed to account for it; with Gauss' discovery, the rather naive point of view that mathematical objects were only "ideas" (in the Platonic sense) of sensory objects became untenable, and gradually gave way to a clearer comprehension of the much greater complexity of the question, wherein it seems to us today that mathematics and reality are almost completely independent, and their contacts more mysterious than ever.

---

* For a clear exposition of this viewpoint, which is due to Hilbert, see the article by C. G. Hempel in the bibliography.
† "L'Oeuvre Mathématique de C. F. Gauss," Poulet-Malassis Alençon: L'Imprimerie Alençonnaise.

The Controversy About the Foundations of Mathematics

It would be misleading to say that mathematics is just a formal game played with symbols and having no broader significance. Mathematicians do not arbitrarily make up axioms—it is unlikely that anyone would ever develop a geometry in which it was assumed that nonadjacent right angles were never congruent to each other. Axioms must lead to interesting and fruitful results. Of course, some axioms that appear uninteresting may turn out to have surprising consequences—this was the case with the hyperbolic axiom, which was virtually ignored during the lifetimes of Gauss, Bolyai, and Lobachevsky. If, however, axiom systems do not bear interesting results, they become neglected and eventually forgotten.

Arguing against the description of mathematics as a "formal game," R. Courant and H. Robbins (in their fine book *What is Mathematics?*) insist that "a serious threat to the very life of science is implied in the assertion that mathematics is nothing but a system of conclusions drawn from definitions and postulates that must be consistent but otherwise may be created by the free will of the mathematician. If this description were accurate, mathematics could not attract any intelligent person. It would be a game with definitions, rules and syllogisms, without motivation or goal."

And Hermann Weyl has remarked: "The constructions of the mathematical mind are at the same time free and necessary. The individual mathematician feels free to define his notions and to set up his axioms as he pleases. But the question is, will he get his fellow mathematicians interested in the constructs of his imagination? We can not help feeling that certain mathematical structures which have evolved through the combined efforts of the mathematical community bear the stamp of a necessity not affected by the accidents of their historical birth."*

Axiom systems that are fruitful can also be controversial in the mathematical world, as are the axioms for infinite sets developed by Georg Cantor, E. Zermelo and others. A controversy occurs because some outstanding mathematicians (such as Weyl, L. E. J. Brouwer, and Errett Bishop in the case of infinite sets) simply do not *believe* all these axioms. If axioms were truly meaningless formal statements, how could there be any controversy about them? Is there any controversy about the

---

* From "A Half-Century of Mathematics," *Amer. Math. Monthly,* **58**: 523–553.

Kurt Gödel

rules of chess? It would seem that the formalist viewpoint—the view that mathematics is just a formal game—is a dodge to avoid having to face the difficult philosophical and psychological problem of the nature of mathematical creations or discoveries. Just what is asserted when a mathematician claims that something exists? When the Pythagoreans discovered that the hypotenuse of an isosceles right triangle was not commensurable with the leg, they tried to keep this discovery secret, calling such lengths "irrational." Nowadays we aren't upset over numbers like $\sqrt{2}$. Similarly, mathematicians have accommodated themselves to "imaginary" numbers, such as $i = \sqrt{-1}$, introduced by J. Cardan.*

The most "fundamentalist" position on the philosophy of mathematics is that of Leopold Kronecker, who dominated the German mathematical world in the late nineteenth century. According to Kronecker, "God created the whole numbers—all else is man-made." This position was later attacked by Hilbert, who vowed that "no one shall expel us from the paradise [of infinite cardinal and ordinal numbers] which Cantor has created for us." The fundamental conceptual changes in Cantor's work were so repulsive to Kronecker that he blocked Cantor from obtaining a professorship at the better German universities and even prevented him from having his papers published in any of the German mathematical journals. Kronecker's attacks against Cantor's work were undoubtedly instrumental in driving Cantor to the insane asylum (see E. T. Bell, *Men of Mathematics*).

Today, Cantor's set theory has been accepted by most mathematicians as the foundation for all of mathematics. For a time, however, mathematicians were uncertain about two axioms in set theory—the axiom of choice and the continuum hypothesis—just as mathematicians earlier were uncertain about the parallel postulate. History repeated itself in 1963 when these axioms were proved independent both of each other and the other axioms of set theory.†

One mystery about mathematics is perhaps the most compelling of all. If mathematical creations are merely arbitrary fancies, how is it that some

---

* Jacques Hadamard has said about Cardan: "It would naturally be expected that the discovery of imaginaries, which seems nearer to madness than to logic and which, in fact, has illuminated the whole mathematical science, would come from such a man whose adventurous life was not always commendable from the moral point of view, and who from childhood suffered from fantastic hallucinations…"

† By Paul J. Cohen; see "Non-Cantorian Set Theory," P. J. Cohen and R. Hersh, *Scientific American*, December, 1967.

turn out to have physical applications, for example, applications that enable us to calculate orbits well enough to put men on the moon? When the Greeks developed the theory of ellipses they had no inkling that it would have applications to a "space race."*

These questions and viewpoints are not intended to confuse you, but to point up the fact that mathematics is alive, ever changing, and incomplete. Moreover, according to a metamathematical theorem of Kurt Gödel, mathematics is forever destined to remain incomplete. He proved that there will always be valid mathematical statements that cannot be demonstrated from systems of axioms that are broad enough to include arithmetic (see DeLong). In other words, Gödel provided a formal demonstration of the inadequacy of formal demonstrations!

Perhaps the following remarks by René Thom† are an appropriate reaction to Gödel's Incompleteness Theorem:

> The mathematician should have the courage of his private convictions; he would then affirm that mathematical structures have an existence independent of the human mind that thinks about them. The form of this existence is undoubtedly different from the concrete and material existence of the external world, but it is nevertheless subtly and profoundly linked to objective existence. For how else explain—if mathematics is merely a gratuitous game, the random product of our cerebral activities—its indisputable success in describing the universe? Mathematics is encountered—not only in the rigid and mysterious laws of physics—but also, in a more hidden but still indubitable manner, in the infinitely playful succession of forms of the animate and inanimate world, in the appearance and destruction of their symmetries. That's why the Platonic hypothesis of Ideas informing the universe is—despite appearances—the most natural and philosophically the most economical. But, at any instant, mathematicians have only an incomplete and fragmentary vision of this world of Ideas . . . , we have to recreate it in our consciousness by a ceaseless and permanent reconstruction. . . . With this confidence in the existence of an ideal universe, the mathematician will not overly worry about the limits of formal procedures, he will be able to forget the problem of consistency. For the world of Ideas infinitely exceeds our operational possibilities, and the ultima ratio of our faith in the truth of a theorem resides in our intuition—a theorem being above all, according to a long-forgotten etymology, the object of a vision.

---

* See E. Wigner, "The Unreasonable Effectiveness of Mathematics," *Communications in Pure and Applied Mathematics,* v. 13 (1960), p. 1ff.

†"'Modern' Mathematics: An Educational and Philosophic Error?" in *American Scientist,* November 1971, p. 695ff. The translation here is my own from the original (in *L'Age de Science,* III (3): 225).

Review Exercise

Which of the following statements are correct?

(1) It is impossible to verify by physical experiments whether hyperbolic geometry is true because hyperbolic geometry is not about physical entities.

(2) If we interpret the undefined terms of geometry physically, e.g., by interpreting "line" as "path of a light ray in empty space," then it makes sense to ask whether this interpretation is a model of Euclidean geometry; however, due to experimental error, physical experiments could never prove conclusively that it is a model.

(3) Hyperbolic geometry is consistent if and only if Euclidean geometry is consistent.

(4) Poincaré maintained that it was meaningless to ask which geometry is "true," and that it only made sense to ask which geometry is more "convenient" for physics.

(5) The most convenient geometry for astrophysics is neither Euclidean nor hyperbolic geometry but a more complicated geometry of space-time developed by Einstein out of ideas from Riemann.

(6) The Klein and Poincaré models, although they appear to be different, are actually isomorphic to each other.

(7) Hyperbolic geometry, although equally as consistent as Euclidean geometry, has no application to other branches of mathematics or to other sciences.

Some Topics for Essays

1. Comment on this quotation from Albert Einstein: "As far as the mathematical theorems refer to reality, they are not sure, and as far as they are sure, they do not refer to reality." (See Hempel for a development of this theme.)

2. Report on the debate about the philosophy of *conventionalism*, using Grünbaum, Poincaré, and Nagel as sources.

3. Report on the use of hyperbolic geometry to describe binocular vision, referring to Luneburg and Blank (see note, p. 250).

4. It can be said that the discovery of non-Euclidean geometry led to the extensive modern development of mathematical logic. Elaborate on this statement, using DeLong, Chapters 1 and 2, as a source.

5. Jacques Hadamard said:* "Practical application is found by not looking for it, and one can say that the whole progress of civilization rests on that principle. . . . It seldom happens that important mathematical researches are *directly* undertaken in view of a given practical use: they are inspired by the desire which is the common motive of every scientific work, the desire to know and understand."

Along the same lines, David Hilbert maintained that in spite of the importance of the applications of mathematics, these must never be made the measure of its value. And the mathematician Jacobi said that "the glory of the human spirit is the sole aim of all science."

Nevertheless, Lobachevsky believed that "there is no branch of mathematics, however abstract, that may not someday be applied to phenomena of the real world."

Comment on these viewpoints.

6. Read the "Socratic Dialogue on Mathematics" in Renyi, and discuss the following questions therein:

(a) "Is it not mysterious that one can know more about things which do not exist than about things which do exist?"

(b) "How do you explain that, as often happens, mathematicians living far from each other and having no contact independently discover the same truths?"

7. Comment on the following statement by Michael Polanyi (in his treatise *Personal Knowledge*; see especially Chapter 6, §9–11):

> We can now turn to the paradox of a mathematics based on a system of axioms which are not regarded as self-evident and indeed cannot be known to be mutually consistent. To apply the utmost ingenuity and the most rigorous care to prove the theorems of logic or mathematics, while the premises of these inferences are cheerfully accepted, without any grounds being given for doing so . . . might seem altogether absurd. It reminds one of the clown who solemnly sets up in the middle of the arena two gateposts with a securely locked gate between them, pulls out a large bunch of keys, and laboriously selects one which opens the lock, then passes through the gate and carefully locks it after himself—while all the while the whole arena lies open on either side of the gateposts where he could go round them unhindered.

---

* *Psychology of Invention in the Mathematical Field*; see especially Chapter 9.

8. Comment on the following statements:

> There is a scientific taste just as there is a literary or artistic one.... Concerning the fruitfulness of the future result—about which, strictly speaking, we most often do not know anything in advance—[the] sense of beauty can inform us and I cannot see anything else allowing us to foresee.... Without knowing anything further we *feel* that such a direction of investigation is worth following.... Everybody is free to call or not to call that a feeling of beauty. This is undoubtedly the way the Greek geometers thought when they investigated the ellipse, because there is no other conceivable way. (Jacques Hadamard, *Psychology of Invention in the Mathematical Field.*)
>
> We dwell on mathematics and affirm its statements for the sake of its intellectual beauty.... For if this passion were extinct, we would cease to understand mathematics; its conceptions would dissolve and its proofs carry no conviction. Mathematics would become pointless and lose itself in a welter of insignificant tautologies.... (Michael Polanyi, *Personal Knowledge.*)

And tell how the following statement by the great number theorist Emil Artin applies or does not apply to your experience with this course: "We all believe that mathematics is an art. The author of a book, the lecturer in a classroom tries to convey the structural beauty of mathematics to his readers, to his listeners. In this attempt he must always fail."

9. Comment on the following statements. G. H. Hardy, in *A Mathematician's Apology,* said:

> For me, and I suppose for most mathematicians, there is another reality, which I will call "mathematical reality"; and there is no sort of argument about the nature of mathematical reality among either mathematicians or philosophers. . . . A man who could give a convincing account of mathematical reality would have solved very many of the most difficult problems of metaphysics. . . . I believe that mathematical reality lies outside us, that our function is to discover or *observe* it, and that the theorems which we prove, and which we describe grandiloquently as our "creations", are simply the notes of our observations. This view has been held, in one form or another, by many philosophers of high reputation from Plato onwards. . . .

Heinrich Hertz, the discoverer of radio waves, said:

> One cannot escape the feeling that these mathematical formulas have an independent existence and an intelligence of their own, that they are wiser than we are, wiser even than their discoverers, that we get more out of them than was originally put into them.

10. Comment on the following remarks by Kurt Gödel:*

I don't see any reason why we should have less confidence in this kind of perception, i.e., in mathematical intuition, than in sense perception, which induces us to build up physical theories and to expect that future sense perceptions will agree with them and, moreover, to believe that a question not decidable now has meaning and may be decided in the future. The set theoretical paradoxes are hardly any more troublesome for mathematics than deceptions of the senses are for physics. . . . Evidently the "given" underlying mathematics is closely related to the abstract elements contained in our empirical ideas. It by no means follows, however, that the data of this second kind [mathematical intuitions], because they cannot be associated with actions of certain things upon our sense organs, are something purely subjective, as Kant asserted. Rather, they too, may represent an aspect of objective reality. But as opposed to the sensations, their presence in us may be due to another kind of relationship between ourselves and reality.

Gödel in this passage speaks primarily of *set theoretical intuition*. As far as geometrical intuition is concerned, the following, according to Gödel,† would have to be added:

Geometrical intuition, strictly speaking, is not mathematical, but rather a priori physical intuition. In its purely mathematical aspect our Euclidean space intuition is perfectly correct, namely, it represents correctly a certain structure existing in the realm of mathematical objects. Even physically it is correct "in the small."

11. Comment on the following quotation from Rolf R. Loehrich:‡

The communication of a new mathematical system or game meets with peculiar obstacles. Each mathematician has a preferred game. A new game may not capture his interest if it is significantly different from those he has been accustomed to play. . . .

A mathematical system is hardly ever presented axiomatized at its inception. Successful axiomatization is a fruition of an *exercitium cogitandi*. Once a system is axiomatized, mathematical activity can be played as a game, as a manipulation of symbols by virtue of rule-systems thought of as invented, but this does not assert that the mathematician who invented or presumably discovered the system meant to play a game. . . . Roberts and I are convinced that there is what might be adequately referred to as a mathematical universe.

---

* From his article "What is Cantor's Continuum Problem?" in Benacerraf and Putnam's *Philosophy of Mathematics,* 2d ed., Prentice-Hall, 1964, p. 271.
† Private communication to the author, October, 1973.
‡ *The Total Field,* with L. G. Roberts, unpublished manuscript.

We believe that, with the complex instrumentations and empirical data set forth in *The Total Field,* the ontological value of confrontations belonging to this universe can be determined with a high degree of accuracy (such confrontations are to be thought [of as] sign-values of signs, and these signs are the symbol systems as known and/or as to be invented by virtue of new conceptual systems with ever increasing ranges).... If this is true, then indeed a mathematician may think of himself as an explorer of the mathematical universe, and any new mathematical system functions as the inception of a possible creation of a universe which comprehends any of the other universes.

12. Write an essay on the development of geometry in ancient Greece, using the resources of your school library. You may be particularly interested in the female mathematician Hypatia.

13. Comment on the following remarks by André Weil, one of the great mathematicians of our time :*

(a) "For us, whose shoulders sag under the weight of the heritage of Greek thought and who walk in the paths traced out by the heroes of the Renaissance, a civilization without mathematics is unthinkable. Like the parallel postulate, the postulate that mathematics will survive has been stripped of its 'evidence'; but, while the former is no longer necessary, we would not be able to get on without the latter."

(b) "Are we witnessing the beginning of a new eclipse of civilization? Rather than to abandon ourselves to the selfish joys of creative work, is it not our duty to put the essential elements of our culture in order, for the mere purpose of preserving it, so that at the dawn of a new Renaissance our descendants may one day find them intact?"

(c) "It is certain that few men of our times are as completely free as the mathematician in the exercise of their intellectual activity. Even if some State ideologies sometimes attack his person, they have never yet presumed to judge his theorems.... The exiled mathematician—and who among us can today feel free from the danger of exile—can find everywhere the modest livelihood which allows him to pursue his work to some extent. Even in jail one can do good mathematics if one's courage fail him not." (Weil did some of his best work while in prison for refusing to serve in the French army.)

* "The Future of Mathematics," *American Mathematical Monthly,* **57** (1950): 295–306.

14. Comment on David Hunter's elaboration of his statement in Exercise 15, Chapter 2:

> Logic is a game invented and played by people who enjoy playing the game of logic. It is very useful in writing books on logic. Excellent too in logic courses. Beyond that what use has it?—so little application to human behavior or the happenings with which human beings are confronted early in the day or at night, waking and dreaming, or slightly sleeping while walking, or smiling with a cup of coffee at the lips, with or without cream and with or without a syllogism in the vest pocket or in the memory. What application has it to the myriad expectations which we impose on tomorrow, on friends, on enemies? All these events and expectations of events happen to happen just as they happen, whenever, in the memory or behind a large tree, after midnight or early in the week, regardless. So!
>
> Of course there is that area in the mind which has been trained to believe we must behave logically. It is constantly prodding us to conform to the imperatives of its syllogisms. But isn't this similar to putting signs along the banks of the Mississippi in April: "Private Property. Do NOT Overflow". Rivers, humans, winds and other imponderables including life often forget to read.
>
> So wear the glasses of logic if you must and you will see the logic in the glasses. You will perhaps assume it is in the events—unless you don't need such glasses. Then you might see something else—ah, yes, indeed, isn't it?

15. Write an essay on a topic of your own.

Appendix | A

Further Results in Hyperbolic Geometry

The theorems of this geometry appear to be paradoxical and, to the uninitiated, absurd; but calm, steady reflection reveals that they contain nothing at all impossible.

C. F. Gauss

Area and Defect

In 1799, in answer to a letter from Wolfgang Bolyai in which Bolyai claimed to have proved Euclid's fifth postulate, Gauss wrote:

> ... the way in which I have proceeded does not lead to the desired goal, the goal that you declare you have reached, but instead to a doubt of the validity of [Euclidean] geometry. I have certainly achieved results which most people would look upon as proof, but which in my eyes prove almost nothing; if, for example, one can prove that there exists a right triangle whose area is greater than any given number, then I am able to establish the entire system of [Euclidean] geometry with complete rigor. Most people would certainly set forth this theorem as an axiom; I do not do so, though certainly it may be possible that, no matter how far apart one chooses the vertices of a triangle, the triangle's area still stays within a finite bound. I am in possession of several theorems of this sort, but none of them satisfy me.*

Perhaps the most surprising fact in J. Bolyai's "strange new universe" is that there is an upper limit to the possible area a triangle can have, even though there is *not* an upper limit to the lengths of the sides of the triangle.

To see how this can be, we have to review the way in which area is calculated in Euclidean geometry. The simplest figure is a rectangle, whose area we calculate as the length of the base times the length of the side.

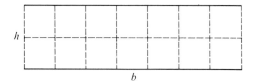

Figure A.1

Area $= bh$.

---

* Bonola, 1955.

This formula is arrived at by noticing that exactly bh unit squares fill up the interior of the rectangle, where a unit square is a square whose side has length one. Keep in mind that the unit of length is arbitrary, so that if we measure area in square inches, we get a different number than if we measure in square feet (but the latter number is always proportional to the former, the proportionality factor being $144 = 12^2$).

From the area of a rectangle we can calculate the area of a right triangle. A diagonal of a rectangle divides it into two congruent right triangles, and since we want congruent triangles to have the same area, the area of the right triangle must be half the area of the rectangle.

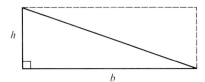

Figure A.2
Area $= \frac{1}{2}bh$.

We can decompose the interior of an arbitrary triangle into the union of the interiors of two right triangles by dropping an appropriate altitude. Since we want the area of the whole to be the sum of the areas of its parts, we find the area of the triangle to be $\frac{1}{2}b_1 h + \frac{1}{2}b_2 h$, and since $b = b_1 + b_2$, we again get half the base times the height for the area of a general triangle.

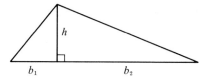

Figure A.3
Area $= \frac{1}{2}bh$ where $b = b_1 + b_2$.

You can see that by taking b and h to be sufficiently large, the area $\frac{1}{2}bh$ can be made as large as you like (e.g., taking b to be 2 million, h 1 million, the area will be 1 trillion square units).

So why doesn't this work equally well in hyperbolic geometry? Because the whole system of measuring area is based on square units, and as we have seen (Theorem 4.1), rectangles (in particular, squares) do not exist in hyperbolic geometry!

What then does "area" mean in hyperbolic geometry? We can certainly say intuitively that it is a way of assigning to every triangle a certain positive number called its *area,* and we want this area function to have the following properties:

1. *Invariance under congruence.* Congruent triangles have the same area.
2. *Additivity.* If a triangle T is split into two triangles T_1 and T_2 by a segment joining a vertex to a point of the opposite side, then the area of T is the sum of the areas of T_1 and T_2.

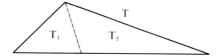

Figure A.4

Area T = Area T_1 + Area T_2.

Having defined area, we then ask how it is calculated. It can be proved rigorously that in hyperbolic geometry the area of a triangle can not be calculated as half the base times the height (see Moise, p. 343). So how do you calculate it? Here we find one of the most beautiful aspects of mathematics, a direct relationship between two concepts that at first seem totally unrelated. You may have recognized this relationship in reading area properties (1) and (2), for in Theorem 4.7 and Exercise 1, Chapter 4, we proved that the *defect* also has these properties. Recall that in hyperbolic geometry the angle sum of any triangle is always less than 180° (Theorem 6.1), so that if we define the defect to be 180° minus the angle sum, we get a positive number. When a mathematician sees two functions with the same properties, he suspects they are related. Gauss discovered this relationship as early as 1794 (he was only 17 years old), and called it the first theorem in the subject.

THEOREM A.1.   In hyperbolic geometry there is a positive constant k such that for any triangle $\triangle ABC$

$$\text{area}\,(\triangle ABC) = \frac{\pi}{180}k^2 \times \text{defect}\,(\triangle ABC).$$

For the proof, which is not difficult although it is somewhat lengthy, see Moise (p. 345). The theorem says that the area of any triangle is proportional to its defect, with proportionality constant $(\pi/180)k^2$. This constant depends on the unit of measurement, i.e., on whichever triangle is taken to have area equal to 1.

We can now see why there is an upper limit to the area of all triangles. Namely, the defect measures how much the angle sum is less than 180. Since the angle sum can never get below 0, the defect can never get above 180. Thus, we have the following corollary.

COROLLARY.   In hyperbolic geometry the area of any triangle is at most πk^2.

Of course, there is no finite triangle whose area equals the maximal value πk^2, although we can approach this area as closely as we wish (and achieve it with an infinite trebly asymptotic triangle). However, J. Bolyai showed how to construct a *circle* of area πk^2 and a *rhombus* (quadrilateral, all of whose angles and sides are congruent) with a 45° angle that also has area πk^2 (see Bonola, pp. 106–110). By the way, the formula for the area of a circle of radius r in hyperbolic geometry is

$$4\pi k^2 \sinh^2 \frac{r}{2k},$$

where $\sinh x = \frac{1}{2}(e^x + e^{-x})$ is the hyperbolic sine of x (Wolfe, p. 169); the circumference of the circle is $2\pi k \sinh(r/k)$ (Wolfe, p. 161).

If we interpret hyperbolic geometry in our physical world, it is clear that since defects of terrestrial triangles are immeasurably small, while

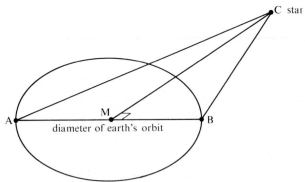

Figure A.5

Defect $(\triangle \text{ACM})° = 90° - (\,\measuredangle\,\text{CAM})° - (\,\measuredangle\,\text{ACM})° < 90° - (\,\measuredangle\,\text{CAM})° = $ parallax.

their areas are measurable, the proportionality constant k^2 must be extremely large. According to Kulczycki (pp. 153–155) the measurements of the parallaxes of fixed stars "elicit that the constant k is not less than about six hundred trillion miles." Thus, $k^2 > 36 \times 10^{28}$. These measurements are based on the fact that the right triangle whose one side is half the major axis of the earth's orbit around the sun and whose opposite vertex is at a fixed star has defect less than the parallax of the star.

The Angle of Parallelism

Recall that given any line l and any point P not on l, there exist limiting parallel rays $\overrightarrow{PX}$ and $\overrightarrow{PY}$ to l that are situated symmetrically about the perpendicular PQ from P to l (see Chapter 6). We proved that $\not\gtrless XPQ \cong \not\gtrless YPQ$ (Theorem 6.6), so either of these angles can be called the *angle of parallelism* for P with respect to l.

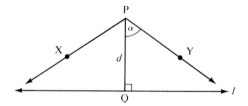

Figure A.6

It is not difficult to show that the number α of radians in the angle of parallelism depends only on the distance d from P to Q, not on the particular line l or the particular point P (see Major Exercise 5, Chapter 6). In the notation of p. 161, $\alpha = (\pi/180)\Pi(PQ)°$.

The following formula relating α and d was discovered by J. Bolyai and Lobachevsky.

THEOREM A.2 (Formula of Bolyai-Lobachevsky).

$$\tan\frac{\alpha}{2} = e^{-d/k},$$

where k is the constant whose square occurs in the proportionality factor of area to defect in Theorem A.1.

This is certainly one of the most remarkable formulas in all of mathematics, and it is astonishing how few mathematicians know it. In this formula the number e is the base for the natural logarithms (e is approximately 2.718 . . .), and tan $\alpha/2$ is the trigonometric tangent of half of α. Proofs of this formula will be found in Kulczycki (§ 20) and in Borsuk and Szmielew (Chapter 6, § 26). We have checked this formula for the Poincaré disk model (where $k = 1$)—see Theorem 7.2, p. 212.

Strange Curves

The various proofs of the Bolyai-Lobachevsky formula (Theorem A.2) all make use of a curve that is peculiar to hyperbolic geometry, called either a *limiting curve* or a *horocycle* in the literature. It is obtained as follows:

Start with a line l, point Q on l, and erect perpendicular $\overleftrightarrow{PQ}$ to l through Q. Then consider the circle δ with center P and radius PQ, which is tangent to l at Q. Now let P recede from l along the perpendicular. The circle δ will increase in size, remaining tangent to l, and will approach

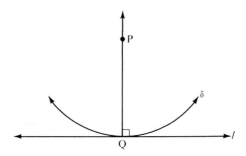

Figure A.7

a limiting position as P recedes arbitrarily far from Q. In Euclidean geometry the limiting position of δ would just be the line l, but in hyperbolic geometry the limiting position of δ is a new curve h called a *limiting curve* or *horocycle*.

We can visualize this in the Poincaré model as follows: Let l be a diameter of the Euclidean circle γ whose interior represents the hyperbolic plane, and let Q be the center of γ. It can be proved that the hyperbolic circle with hyperbolic center P is represented by a Euclidean circle whose Euclidean center R lies between P and Q.

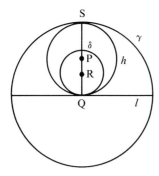

As P recedes from Q towards the ideal point represented by S, R is pulled up to the Euclidean midpoint of SQ, so that the horocycle *h* is a Euclidean circle tangent to γ at S and tangent to *l* at Q. It can be shown in general that all horocycles are represented in the Poincaré model by Euclidean circles inside γ and tangent to γ. Moreover, all the Poincaré lines passing through the ideal point S are orthogonal to *h*; a hyperbolic ray from a point of *h* out to the ideal point S is called a *diameter* of *h* (see Wolfe, pp. 213–214).

In the Poincaré model two horocycles tangent to γ at S are said to be *concentric*. It is by studying the ratio of corresponding arcs on concentric horocycles that the Bolyai-Lobachevsky formula can be derived (see Kulczycki, § 18).

There is an analogous construction in hyperbolic space called the *horosphere*. Instead of taking the limit of circles to get the horocycle, one takes the limit of spheres to get the horosphere or *limiting surface*.

Another strange curve in hyperbolic geometry that has no Euclidean

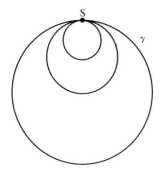

Figure A.9
Concentric horocycles.

counterpart is the *equidistant curve.* Start with a line *l* and a point P not on *l*. Consider the locus of all points on the same side of *l* as P and at the same perpendicular distance from *l* as P. In Euclidean geometry this locus would just be the unique line through P parallel to *l,* but in hyperbolic geometry it is not a line, it is the *hypercycle,* or *equidistant curve,* through P.

In the Poincaré model let A and B be the ideal end points of *l*. It turns out that the equidistant curve to *l* through P is represented by the arc of the Euclidean circle passing through A, B, and P (Wolfe, p. 214).

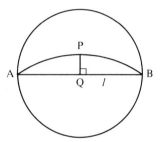

Figure A.10
Equidistant curve.

This curve is orthogonal to all Poincaré lines perpendicular to the line *l*.

In the Euclidean plane three points either lie on a uniquely determined line or on a uniquely determined circle. Not so in the hyperbolic plane— look at the Poincaré model. A Euclidean circle represents:

(a) a *hyperbolic circle* if it is entirely inside γ;
(b) a *horocycle* if it is inside γ except for one point where it is tangent to γ;
(c) an *equidistant curve* if it cuts γ nonorthogonally in two points;
(d) a *hyperbolic line* if it cuts γ orthogonally.

Thus, we can conclude that in the hyperbolic plane three points lie either on a line, a circle, a horocycle, or an equidistant curve, which is uniquely determined.

We previously described the horocycle as a limit of circles (Figure A.7). It can also be described as a limit of equidistant curves. This time in Figure A.10 fix P and move Poincaré line *l* away from P, keeping it perpendicular to $\overleftrightarrow{PQ}$. The limiting position of the equidistant curves obtained from this movement will be the horocycle through P centered at the ideal endpoint of $\overrightarrow{PQ}$.

The Pseudosphere

One of the difficulties with the Poincaré model is that, although it faithfully represents angles of the hyperbolic plane (i.e., it is a *conformal model*), it distorts distances. So it is natural to ask whether another model exists that also represents hyperbolic lengths faithfully by Euclidean lengths. If there is such a model, it would be called *isometric*. An equally natural idea is to seek as a model some surface in Euclidean three-dimensional space. The lines of the hyperbolic plane would then be represented by *geodesics* on the surface, and we would expect the surface to be curved so as to mirror our expectation that hyperbolic lines are "really curved." (By definition, the *geodesic segment* between two points on a surface is the shortest path lying on the surface between those points, e.g., on the surface of a sphere, a geodesic segment is an arc of a great circle.)

A difficult theorem of Hilbert states that it is impossible to embed the entire hyperbolic plane isometrically as a surface in Euclidean space. On the contrary, it *is* possible to embed the Euclidean plane isometrically in hyperbolic space, as the surface of the horosphere (see Kulczycki, § 17). This result, proved by both J. Bolyai and Lobachevsky, was already recognized by Wachter in 1816.*

But all is not lost. It turns out to be possible to embed a portion of the hyperbolic plane isometrically in Euclidean space, the portion called a *horocyclic sector,* bounded by an arc of a horocycle and the two diameters cutting off this arc. Such a sector in the Poincaré model is shown in Figure A.11 :

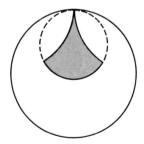

Figure A.11

---

* For those who understand this language, strangely enough there does exist a continuously differentiable embedding of the hyperbolic plane into Euclidean 3-space. This was proved in 1955 by N. Kuiper, using analytic methods (see *Indagationes Mathematicae,* **17**: 683). It is known that no "nice," e.g., C^2, embeddings exist.

The surface that represents this region isometrically is called a *pseudosphere*. It is obtained by rotating a curve called a *tractrix* around its asymptote. It looks like an infinitely long horn. (In this representation the two diameters of the horocyclic arc have been identified (pasted together) so that the mapping of the sector into Euclidean space is actually only an embedding of the region between the diameters.) The tractrix is characterized by the fact that the tangent from any point on the curve to the vertical asymptote has constant length *a*.* The representation on the pseudosphere was discovered by Beltrami, using previous work by F. Minding.

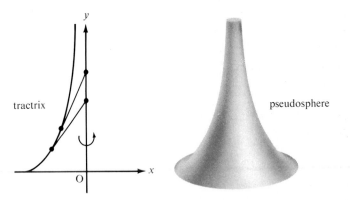

Figure A.12

The representation on the pseudosphere enables us to give some geometric meaning to the fundamental constant *k* that appears in Theorems A.1 and A.2. The point is that there is a way (discovered by Gauss) of measuring the *curvature* of any surface. We cannot give the precise definition, since it involves a knowledge of differential geometry (see Coxeter's *Introduction to Geometry*; Hilbert and Cohn-Vossen; or Appendix B of this book). Intuitively, however, it is a number *K* that measures the extent to which the surface "bends." In general, the curvature *K* varies from point to point, being close to zero at points where the surface is rather flat, large at points where the surface bends sharply. For some surfaces the curvature is the same at all points, so naturally these are called *surfaces of constant curvature K*.

* The Dutch physicist Huygens called the tractrix the "dog curve" because it resembles the curve described by the nose of a dog being dragged reluctantly on a leash.

For a surface of constant curvature K, Gauss proved a fundamental formula relating the curvature, area, and angular measure. He took a geodesic triangle $\triangle ABC$ with vertices A, B, and C and sides geodesic segments. By integration, he calculated the area of the triangle. He determined that, if $(\angle A)^r$ denotes the radian measure of angle A, then

$$K \times \text{area } \triangle ABC = (\angle A)^r + (\angle B)^r + (\angle C)^r - \pi.$$

(Recall that $(\angle A)^r = \pi/180(\angle A)^\circ$.) He then showed what this meant by considering the three possible cases:

Case 1. K is positive, hence, both sides of the equation are positive. In this case Gauss' formula shows that the angle sum in radians of a geodesic triangle is greater than π (or in degrees, greater than 180°), and that the area is proportional to the *excess* (the number on the right side of the equation), with a proportionality factor of $1/K$. An example could be the surface of a sphere of *radius r*, whose curvature is $K = 1/r^2$. The larger the radius, the smaller the curvature, and the more the surface resembles a plane. (Gauss' formula in the special case of a sphere was already discovered by Girard in the seventeenth century.) According to a theorem of H. Liebmann, H. Hopf and W. Rinow, spheres are the only complete surfaces of constant positive curvature in Euclidean three-space, so that the elliptic plane cannot be embedded in Euclidean three-space either.

Case 2. $K = 0$. In this case Gauss' formula shows that the angle sum in radians is equal to π. An example would be the Euclidean plane; another example is an infinitely long cylinder.

Case 3. K is negative. In this case Gauss' formula shows that the angle sum in radians is less than π, and the area is proportional to the *defect*. An example of such a surface is the *pseudosphere*. Since the pseudosphere represents a portion of the hyperbolic plane isometrically, we can compare Gauss' formula with the formula in Theorem A.1 relating area to defect. The comparison gives $K = -1/k^2$. Thus, $-1/k^2$ is *the curvature of the hyperbolic plane*.

Here we can recapture the analogy with Case 1 by setting $r = ik$, where $i = \sqrt{-1}$. Then $K = 1/r^2$, so that the hyperbolic plane can be described as a "sphere of imaginary radius $r = ik$," as Lambert noticed. This description is not as ridiculous as it seems, for all the formulas of hyperbolic trigonometry can be obtained from the formulas known for spherical trigonometry by replacing r with ik.*

Finally, notice that as k gets very large, the curvature K approaches zero, and the geometry of the surface resembles more and more the geometry of the Euclidean plane. It is in this sense that Euclidean geometry is a "limiting case" of hyperbolic geometry.

---

* There is a mathematical explanation for why this works. The hyperbolic plane of curvature $-1/k^2$ is the "dual symmetric space" to the sphere of radius k, in the sense of Élie Cartan (see S. Helgason, *Differential Geometry and Symmetric Spaces*, p. 206).

Appendix | B

Elliptic and Other Riemannian Geometries

The dissertation submitted by Herr Riemann offers convincing evidence...of a creative, active, truly mathematical mind, and of a gloriously fertile imagination.

C. F. Gauss

Elliptic Geometry

In Euclidean geometry there is exactly one parallel to a line l through a point P not on l; in hyperbolic geometry there is more than one parallel. A third geometry could be studied, one in which there is *no* parallel to l through P, i.e., a geometry in which parallel lines do not exist.

However, if we simply add the latter as a new parallel axiom to replace the other parallel axioms, the system we get is inconsistent. In Corollary 2 to Theorem 4.1 we proved that parallel lines do exist in neutral geometry, so that we would get a contradiction by adding such a parallel axiom.

To avoid this, we have to modify some of our other axioms. We can see what modifications need to be made by thinking of the surface of a sphere and interpreting "line" as "great circle." Then, indeed, there are no parallel lines. But other things change as well. It is impossible to talk about one point B being "between" two other points A and C on a circle. So all the axioms of betweenness have to be scrapped. They are replaced instead by seven axioms of *separation*. In Figure B.1 A and C separate B and D on the circle, since you can't get from B to D without crossing either A or C.

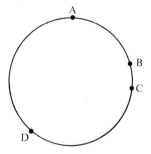

Figure B.1

Let us designate the undefined relation "A and C separate B and D" by the symbol $(A, C|B, D)$. The separation axioms are then:

SEPARATION AXIOM 1. If $(A, B|C, D)$, then points A, B, C, and D are collinear and distinct.

SEPARATION AXIOM 2. If $(A, B|C, D)$, then $(C, D|A, B)$ and $(B, A|C, D)$.

SEPARATION AXIOM 3. If $(A, B|C, D)$, then not $(A, C|B, D)$.

SEPARATION AXIOM 4. If points A, B, C, and D are collinear and distinct, then $(A, B|C, D)$ or $(A, C|B, D)$ or $(A, D|B, C)$.

SEPARATION AXIOM 5. If points A, B, and C are collinear and distinct, then there exists a point D such that $(A, B|C, D)$.

SEPARATION AXIOM 6. For any five distinct collinear points A, B, C, D, and E, if $(A, B|D, E)$, then either $(A, B|C, D)$ or $(A, B|C, E)$.

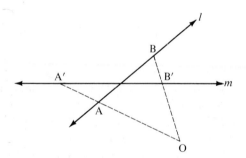

Figure B.2

To state the last axiom, we need the notion of a *perspectivity* from one line onto another (see Chapter 7, p. 222). Let l and m be any two lines and O a point not on either of them. For each point A on l the line $\overleftrightarrow{OA}$ intersects m in a unique point A' (remember the elliptic parallel property); the one-to-one correspondence that assigns A' to A for each A on l is called the *perspectivity* from l to m with center O.

SEPARATION AXIOM 7.   Perspectivities preserve separation, i.e., if
(A, B|C, D), with *l* the line through A, B, C, and D, and if A', B', C',
and D' are the corresponding points on line *m* under a perspectivity,
then (A', B'|C', D').

Without the notion of betweenness we have to carefully reformulate
all the geometry using that relation. For example, the *segment* AB
consists of the points A and B and all points between them. Yet this
doesn't make sense on a circle. We can only talk about the segment
ABC determined by *three* collinear points: it consists of the points A,
B and C and all the points not separated from B by A and C.

Similarly, we have to redefine the notion of a triangle, since its sides
are no longer determined by the three vertices.

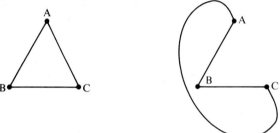

Figure B.3

Two different "triangles" with the same vertices.

Once these notions have been redefined, the axioms of congruence
and continuity all make sense and can be left intact.

There is still a difficulty with Incidence Axiom 1, which asserts that
two points do not lie on more than one line. This is false for great
circles on the sphere, since antipodal points (such as the poles) lie on
infinitely many lines.

Klein saw that the way to remedy this is to *identify* antipodal points,
i.e., just as we interpret "line" to mean "great circle" in this model, we
interpret "point" to mean "pair of antipodal points." This means that *in
our imagination* we have pasted together two antipodal points so that
they coalesce into a single point. It can be proved, as you might guess,
that such pasting cannot actually be carried out in Euclidean three-

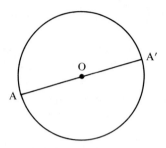

Figure B.4
A and A′ are identified.

dimensional space. But we can still identify antipodal points in our minds—every time we move from one to the other we think of ourselves as being back at the original point.

In making these identifications we discover another surprising property: a line no longer divides the plane into two sides, for you can "jump across" a great circle by passing from a given point to its now equal antipodal point that used to be on the other side. If we cut out a strip from this plane, it will look like a Möbius strip, which has only one side (see Figure B.5). The technical name for this property of "one-sidedness" is *nonorientability*.

To sum up, the axioms of plane elliptic geometry consist of the same incidence, congruence, and continuity axioms as neutral geometry (with the new definitions of segment, triangle, etc.). The betweenness axioms are replaced by separation axioms, and the parallel postulate is replaced by an axiom stating that no two lines are parallel. The model, which shows that elliptic geometry is just as consistent as Euclidean geometry,

Figure B.5
Möbius strip.

consists of the great circles on the sphere with antipodal points identified.*

As you might expect from this model, it is a theorem in elliptic geometry that *lines have finite length.* Moreover, all the lines perpendicular to a line *l* are not parallel to each other but are concurrent, i.e., all the perpendiculars to *l* have a point in common called *the pole of l.* In the model, for instance, the pole of the equator is the north (or, what is the same, the south) pole.

Another model for plane elliptic geometry (due to Klein) is *conformal*— like the Poincaré model for hyperbolic geometry, angles are accurately represented by Euclidean angles. In this model "points" are the Euclidean points inside the unit circle in the Euclidean plane as well as pairs of antipodal points on the circle; "lines" are either diameters of the unit circle or arcs of Euclidean circles that meet the unit circle at the ends of a diameter (see H. M. S. Coxeter's *Non-Euclidean Geometry*, §14.6). This representation shows clearly that the angle sum of a triangle is greater than 180° in elliptic geometry (see Figure B.6).

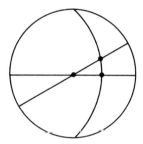

Figure B.6

Elliptic geometry becomes even more interesting when you pass from two to three dimensions. In three dimensions orientability is restored and a new kind of parallelism occurs. Two lines are called *Clifford-parallel* if they are equidistant from each other; the lines are joined to each other by a continuous family of common perpendicular segments of the same length. Such lines cannot lie in a plane (in an elliptic plane two lines must intersect), so they are skew lines. Moreover, in general, in elliptic space for any point P not on a line *l* there exist exactly two lines through

---

* The geometry of the sphere itself is sometimes misleadingly called "double elliptic geometry." For an axiomatic treatment of elliptic geometry, see Borsuk and Szmielew.

P that are Clifford-parallel to *l*, called the *right and left Clifford parallels* to *l* through P. We say "in general" because there is a special line *l*∗, called the *absolute polar* of *l*: if P lies on *l*∗, there is only one Clifford parallel to *l* through P, which is *l*∗. Every plane through *l* is perpendicular to every plane through *l*∗. (Naturally, this is difficult to visualize! See H. M. S. Coxeter, *Non-Euclidean Geometry,* Chapter 7.)

Elliptic space is finite but unbounded—finite because all lines have finite length and look like circles, and unbounded because there is no boundary, just as on the surface of a sphere there is no boundary. In a universe having this geometry, with light rays traveling along elliptic lines, you could conceivably look through a very powerful telescope and see the back of your own head! (Although you might have to wait a few billion years for the light to travel all the way around.)

Riemannian Geometry

It is impossible to rigorously explain the ideas of Riemannian geometry without using the language of the differential and integral calculus, so in this appendix we can only attempt to understand very roughly the intuitive idea. The basic notion we shall attempt to grasp is *curvature*.

The two simplest smooth one-dimensional figures in the Euclidean plane are a line and a circle. We think of a line as being "straight," i.e., not curved, so that if we were to assign any numerical curvature k to a line, we would assign the value $k = 0$. A circle γ, on the other hand, is certainly "curved," and how much it is curved depends on its radius r. The larger the radius of γ, the more γ approaches a line, i.e., the less it curves; it is therefore natural to define the curvature k of a circle of radius r by $k = 1/r$ (the curvature is inversely proportional to the radius).

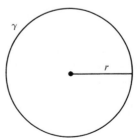

Figure B.7

Curvature $k = \dfrac{1}{r}$.

Consider next an arbitrary smooth curve γ in the Euclidean plane. By "smooth curve" we mean one that has a continuously turning tangent line at each point. The tangent to a point on a circle is the perpendicular to the radius at that point. We can define the tangent to a point P on general curve γ as follows: P and another point Q on γ determine a line l. We fix point P and let Q approach P. The limiting position achieved by line l as Q approaches P is by definition the *tangent line t* to γ at P.

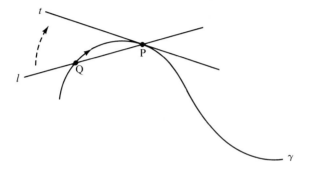

Figure B.8

Besides the fixed point P, we can also consider two other points P_1 and P_2 on γ. These three points determine a circle δ. Fix P and let P_1 and P_2 both approach P along γ. The limiting position of circle δ as P_1 and P_2 approach P is the circle that "best fits" the curve γ at P, and is called the *osculating circle* to γ at P (from the Latin *osculari*, "to kiss"). It is reasonable to define the *curvature* of γ at P as the curvature of its osculating circle at P, i.e., the reciprocal $k = 1/r$ of the radius r of the osculating circle (r is also called the *radius of curvature* of γ at P).

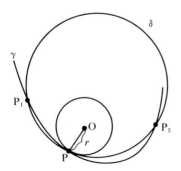

Figure B.9

It is clear from Figure B.10 that the osculating circle will vary in size as we move along the curve γ, so that the curvature k varies from point to point along γ. Notice also that the tangent to γ at a point P is also the tangent to the osculating circle at P.

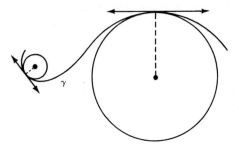

Figure B.10

Notice also, in Figure B.10 that the osculating circle may be on different sides of the curve γ. It is convenient to *redefine curvature* so that it is positive on one side of γ and negative on the other. Once this is done, it becomes clear that in Figure B.10, there must be a point I between A and B on γ at which the curvature is zero, since we assume γ is smooth enough for the curvature to vary continuously. This point I is called a *point of inflection,* and at such a point the osculating "circle" degenerates into a line, the tangent line at I (see Figure B.11).

We next pass to smooth surfaces embedded in Euclidean three-dimensional space. "Smooth" now means that there is a continuously turning *tangent plane* for each point P of the surface. Consider the line through P that is perpendicular to the tangent plane, called the *normal*

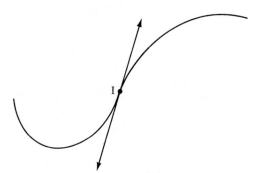

Figure B.11
Point of inflection $k = 0$.

line at P. A plane that contains the normal line intersects the surface in a plane curve. We can imagine this plane rotating around the normal line, and as it does so, it will cut out different curves on the surface passing through P. We have already explained how to define the curvature at P for each of these normal sections. In general, these curvatures will vary as we rotate around the normal line. (In the special case of a sphere these curvatures are constant and equal to the reciprocal of the radius of the sphere, since the curves cut out are all great circles on the sphere.) It can be proved by methods of differential geometry that these curvatures achieve a maximum value k_1 and a minimum value k_2 as we rotate, and the corresponding normal sections (called *principal curves*) are perpendicular to each other. The product $K = k_1 k_2$ of these maximum and minimum curvatures is now called the *Gaussian curvature* (after Gauss, who first studied it), or simply "the curvature," of the surface at the point P. Once again, K will in general change as P varies over the surface; if K happens to stay constant, we obtain the three geometries discussed in Appendix A, according as K is negative (pseudosphere), zero (plane), or positive (sphere).

Figure B.12

In Figure B.12, the tangent plane, normal line, and principal curves for a saddle-shaped surface are shown. For point P on this surface the Gaussian curvature will be a negative number, according to our convention, since the osculating circles for the two principal curves lie on different sides of the tangent plane. On the other hand, for the egg-shaped surface in Figure B.13, the two principal curves lie on the same side of the tangent plane, so the Gaussian curvature is positive.

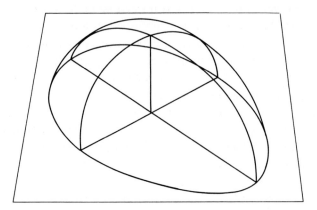

Figure B.13

In the case of a cylinder one of the principal curves will be a straight line, which has curvature $k_2 = 0$; hence, the Gaussian curvature $K = k_1 k_2$ at any point on a cylinder will also be zero (see Figure B.14). We can better grasp this surprising result if we think of a cylinder as a "rolled-up plane." Surely, in any sensible definition of surface curvature a flat plane should be assigned zero curvature. In the process of "rolling-up" a rectangular plane strip the arc-lengths and angles between curves on the strip are not changed, and in this sense the "intrinsic geometry" is not changed. Gauss was looking for a definition of surface curvature that depended only on the intrinsic geometry of the surface, not on the particular way the surface was embedded in Euclidean three-space. He was able to prove that his curvature K did not change if the surface was subjected to a "bending" in which arc-lengths and angles of all curves on the surface are left invariant. Thus, K describes the intrinsic curvature of the surface independent of the way it is bent to fit into Euclidean three-space. This is all the more remarkable because the principal curvatures k_1 and k_2 may change under such a "bending"; nevertheless, their product $K = k_1 k_2$ stays the same. Gauss was so excited about this result that he named it the *theorema egregrium,* "the extraordinary theorem." In a letter to the astronomer Hansen he wrote: "These investigations deeply affect many other things; I would go so far as to say they are involved in the metaphysics of the geometry of space."

Gauss also solved the problem of determining this intrinsic curvature K without reference to the ambient three-space. Imagine a two-dimensional creature living on a surface and having no conception of a third dimension, being unable to conceive of the normal lines we used to define the curvature K. How could this creature calculate K? We will have to use the language of the differential calculus to give Gauss' answer.

In the Euclidean plane a point is determined by its x and y coordinates. If these coordinates are subjected to infinitesimal changes denoted dx and dy, then the point moves an infinitesimal distance ds whose square is given by the Pythagorean formula $ds^2 = dx^2 + dy^2$. Now on a smooth surface a point will also be determined locally by two coordinates x and y. If these coordinates are subjected to infinitesimal changes dx and dy, then the point moves a distance ds on the surface whose square is given by the more complicated expression

$$ds^2 = E\,dx^2 + 2F\,dx\,dy + G\,dy^2,$$

where E, F, and G may vary as the point varies. The functions E, F and G could in principle be determined by the two-dimensional creature

Figure B.14

Gaussian curvature of a cylinder is zero.

Georg Friedrich Bernhard Riemann

making measurements on his surface. Gauss then showed that his curvature K is given in terms of E, F and G by a not too complicated formula (see Lanczos, p. 183). Thus, the creature could also calculate K from this formula and discover that his world was curved, although he would have difficulty visualizing what that might mean.

Although this talk about a two-dimensional creature may seem bizarre, it is not, as Riemann demonstrated. Riemann reasoned that we are in an entirely analogous situation, living in a three-dimensional universe in which an infinitesimal change in distance ds is given by an analogous formula involving the three infinitesimals dx, dy, and dz:

$$ds^2 = g_{11}\,dx^2 + g_{22}\,dy^2 + g_{33}\,dz^2 + 2g_{23}\,dy\,dz + 2g_{31}\,dz\,dx + 2g_{12}\,dx\,dy.$$

From this formula Riemann was able to define a "curvature tensor" analogous to the Gaussian curvature for a surface, only more complicated: Gauss' curvature involved only a single number K, whereas Riemann's depended on six different numbers. Riemann discovered this curvature almost accidentally in his research on heat transfer. In fact, he developed such a curvature tensor for abstract geometries of any dimension n, and Einstein was able to apply Riemann's ideas to his four-dimensional space-time continuum.

So we are in the same position as that poor two-dimensional creature. We can make measurements to determine the Riemannian curvature of our universe. Astronomers have been performing such measurements. If we find that the Riemannian curvature is not zero, we know that the geometry is not Euclidean. However, this does not mean that our space is embedded in some higher dimensional physical space in which it is somehow "curved." When we say, loosely, that "space is curved," we mean only that its geometric properties differ from the properties of Euclidean space in a very specific way given by Riemann's formulas.

It was in his 1854 inaugural lecture, *Ueber die Hypothesen welche der Geometrie zugrunde liegen* ("On the hypotheses that form the foundation for geometry") that Riemann introduced the idea of an n-dimensional space whose intrinsic geometry is determined by a quadratic formula for the infinitesimal change in distance ds. Such a structure is now called a *Riemannian manifold* (see Spivak, 1970, for the precise definition). Different manifolds yield different geometries, so that Riemann discovered an infinite number of new geometries. Each such geometry has as its "lines" the geodesics on the manifold (paths that minimize arc lengths), and

Riemann recognized that the geodesics would be of finite length if the curvature of the manifold was positive.

It is important to note that, although we have emphasized Riemann's generalization of Gauss' ideas from two dimensions to higher dimensions, Riemann's formulation gives new information about *surfaces* that cannot be embedded in Euclidean three-space. For example, the hyperbolic plane can be described as a complete two-dimensional Riemannian manifold of constant negative curvature, and the elliptic plane can be described as a complete two-dimensional Riemannian manifold of constant positive curvature on which two points lie on a unique geodesic. Neither of these manifolds can be analytically embedded in Euclidean three-space.

Some idea of Riemann's influence on modern mathematics can be gleaned from the following list of concepts, methods, and theorems that have been named after him: Riemannian curvature of Riemannian manifolds, Riemann integral, Riemann-Lebesgue lemma, Riemann surfaces, Riemann-Roch theorem, Riemann matrices, Riemann hypothesis about the Riemann zeta function, Riemann's method in the theory of trigonometrical series, Riemann's method for hyperbolic partial differential equations, Riemann mapping theorem, and Cauchy-Riemann equations.*

---

* For the story of Riemann's difficult life, see Bell, *Men of Mathematics*; for a mathematically precise treatment of Riemannian geometry, including an explanation of what Riemann said in his famous 1854 inaugural dissertation lecture, see Spivak's *Differential Geometry*, Volume 2; for applications to the theory of relativity, see Lanczos; for an intuitive discussion of how gravity can be explained in terms of the curvature of space, see the last part of Taylor and Wheeler, *Spacetime Physics*.

Suggestions
For Further Reading

1. If you want to fill in some of the gaps in this book, see Borsuk and Szmielew, where it is proved that the axioms for Euclidean and hyperbolic geometries are categorical and where elliptic geometry is also developed.

2. If you want to know more about the history of non-Euclidean geometry, and perhaps look at the original papers by J. Bolyai and Lobachevsky, see Bonola. For a biting critical account of the confused reaction to non-Euclidean geometry by late nineteenth-century mathematicians, see Freudenthal. See Dodgson for a belated substitute for the parallel postulate—amusingly presented.

3. You may want to know more about other kinds of geometries, such as projective geometry, the study of which will illuminate the mysterious cross-ratio used to define length in the Klein and Poincaré models (Chapter 7). Projective geometry, as an independent science, blossomed in the first half of the nineteenth century in the work of J. V. Poncelet, J. Steiner, J. Plücker, M. Chasles, and C. G. Von Staudt. In the latter half of that century, Klein and Arthur Cayley (1821–1895, England) showed that projective geometry plays a dominant role in the classification of all the other geometries. (For example, elliptic geometry is just real projective geometry with congruence added.) For introductory expositions see Tuller; Aleksandrov; Coxeter (*Projective Geometry*); and Kline's article in *Scientific American*. For advanced treatments see Klein; Coxeter (*Non-Euclidean Geometry*); or Cartan.

4. If you want to go further into hyperbolic geometry, see Wolfe. Since hyperbolic geometry is just a very special part of differential geometry, you should (if you have mathematical ability) master the differential and integral calculus first, then plow into Spivak.

5. You may be interested in another system of axioms for geometries based on the use of motions (such as the reflections, rotations, translations, etc., studied in the M-exercises of Chapter 7). This approach makes

rigorous Euclid's idea of superimposing one figure upon another and is the most modern viewpoint. You should learn some group theory, complex variables, and linear algebra before attempting this. See Bachmann, Ewald, Artzy, Schwerdtfeger and Redei. Using axioms for motions instead of for congruence, Redei has carried out the program sketched by Klein and Sophus Lie for putting Euclidean, hyperbolic, and elliptic geometries on a common axiomatic foundation. For the theory of *H-planes,* based on all our axioms for neutral geometry except the continuity axioms, see Hessenberg and Diller (where non-Archimedean geometries are discussed).

6. If you are curious about the application of non-Euclidean geometry to general relativity and cosmology, see Lanczos or Taylor and Wheeler, then consult a friendly physicist for advice on the vast literature in this field.

Bibliography

Introductory and General

Aleksandrov, A. 1969. "Abstract Spaces," in vol. 3 of *Mathematics: Its Content, Methods and Meaning.* A. Aleksandrov *et al.,* eds., 2nd ed., Cambridge (Mass.): MIT Press.

Courant, R. and H. Robbins. 1941. *What is Mathematics?* Oxford University Press.

Eves, H. and C. Newsom. 1965. *An Introduction to the Foundations and Fundamental Concepts of Mathematics,* rev. ed., New York: Holt, Rinehart, Winston.

Gamow, G. 1956. "The Evolutionary Universe." *Scientific American,* **195**: 136–154. (Offprint No. 211.)

Hilbert, D. and S. Cohn-Vossen. 1952. *Geometry and the Imagination,* New York: Chelsea.

Kline, M. 1962. *Mathematics: A Cultural Approach,* Reading (Mass.): Addison-Wesley.

——, comp. 1968. *Mathematics in the Modern World: Readings from Scientific American,* San Francisco: W. H. Freeman. (See especially the articles on geometry by Kline and Le Corbeiller.)

Wilder, R. L. 1952. *Introduction to the Foundations of Mathematics,* New York: Wiley.

History and Biography

Bell, E. T. 1934. *Search for Truth,* New York: Reynal and Hitchcock.

——. 1961. *Men of Mathematics,* Simon and Schuster.

——. 1969. "Father and Son, Wolfgang and Johann Bolyai," in *Memorable Personalities in Mathematics: Nineteenth Century,* Stanford (Calif.): School Mathematics Study Group.

Bonola, R. 1955. *Non-Euclidean Geometry,* New York: Dover.

Boyer, C. B. 1968. *A History of Mathematics,* New York: Wiley.

Dodgson, C. L. (Lewis Carroll). 1890. *Curiosa Mathematica: A New Theory of Parallels,* London: Macmillan.

Dunnington, G. W. 1955. *Carl Friedrich Gauss: Titan of Science,* New York: Hafner.

Engel, F. and P. Stäckel. 1895. *Theorie der Parallellinien von Euklid bis auf Gauss,* Leipzig: Teubner.

Freudenthal, H. 1962. "The main trends in the foundations of geometry in the nineteenth century," in *Logic, Methodology and Philosophy of Science,* E. Nagel, P. Suppes and A. Tarski, eds., Stanford University Press, 1962, pp. 613–621.

Hall, T. 1970, *C. F. Gauss: A Biography,* MIT Press.

Heath, T. L. 1956. *Euclid's Elements,* New York: Dover.

Lanczos, C. 1970. *Space Through the Ages,* New York: Academic Press.

Nagel, E. 1939. "Formation of Modern Conceptions of Formal Logic in the Development of Geometry," *Osiris,* No. 7, pp. 142–224.

Reid, C. 1970. *Hilbert,* New York: Springer-Verlag.

Saccheri, G. 1970. *Euclides ab omni naevo vindicatus,* tr. G. B. Halsted, New York: Chelsea.

Schmidt, F. and P. Stäckel. 1972. *Briefwechsel Zwischen C. F. Gauss und W. Bolyai,* New York: Johnson Reprint Corp.

Sjöstedt, C. E. 1968. *Le axiome de paralleles de Euclides a Hilbert,* Stockholm: Natur och Kultur.

Stäckel, P. 1913. *Wolfgang und Johann Bolyai, Geometrische Untersuchungen,* 2 vols., Leipzig: Teubner.

Van der Waerden, B. L. 1961. *Science Awakening,* Oxford University Press.

Philosophical

DeLong, H. 1970. *A Profile of Mathematical Logic,* Reading (Mass.): Addison-Wesley.

Grünbaum, A. 1968. *Geometry and Chronometry in Philosophical Perspective,* University of Minnesota Press.

Hadamard, J. 1945. *The Psychology of Invention in the Mathematical Field,* Princeton University Press.

Hardy, G. H. 1940. *A Mathematician's Apology,* Cambridge University Press.

Hempel, C. G. 1945. "Geometry and Empirical Science," *American Mathematical Monthly,* **52**: 7–17; also in vol. 3, *World of Mathematics,* J. R. Newman, ed., pp. 1619–1646.

Nagel, E. 1961. *The Structure of Science,* New York: Harcourt, Brace & World.

Poincaré, H. 1952. *Science and Hypothesis,* New York: Dover.

Polanyi, M. 1964. *Personal Knowledge,* New York: Harper and Row.

Renyi, A. 1967. *Dialogues on Mathematics,* San Francisco: Holden-Day.

Mathematically Elementary to Moderate

Adler, I. 1966. *A New Look at Geometry,* New York: John Day Co.

Artzy, R. 1965. *Linear Geometry,* Reading (Mass.): Addison-Wesley.

Beck, A., M. Bleicher, and D. Crowe. 1969. *Excursions into Mathematics,* New York: Worth.

Coxeter, H. S. M. 1964. *Projective Geometry,* New York: Blaisdell.

———. 1969. *Introduction to Geometry,* New York: Wiley.

Coxeter, H. S. M. and S. L. Greitzer. 1967. *Geometry Revisited,* New York: Random House.

Eves, H. (1963–1965). *A Survey of Geometry,* 2 vols., Boston: Allyn and Bacon.

Golos, E. B. 1968. *Foundations of Euclidean and Non-Euclidean Geometry,* New York: Holt, Rinehart and Winston.

Hessenberg, G., and J. Diller. 1967. *Grundlagen der Geometrie,* Berlin: de Gruyter.

Kay, D. C. 1969. *College Geometry,* New York: Holt, Rinehart and Winston.

Klein, F. 1948. *Geometry,* Part 2 of *Elementary Mathematics from an Advanced Standpoint,* New York: Dover.

Kulczycki, S. 1961. *Non-Euclidean Geometry,* Elmsford (N.Y.): Pergamon.

Kutuzov, B. V. 1960. *Geometry,* Stanford (Calif.): School Mathematics Study Group.

Luneburg, R. K. 1947. *Mathematical Analysis of Binocular Vision,* Princeton University Press.

Meschkowski, H. 1964. *Noneuclidean Geometry,* New York: Academic Press.

Norden, A. P. 1958. *Elementare Einführung in die Lobatchewskische Geometrie,* Berlin: Deutscher Verlag der Wissenschaften.

Pedoe, D. 1970. *A Course of Geometry,* Cambridge University Press.

Prenowitz, W. and M. Jordan. 1965. *Basic Concepts of Geometry,* New York: Blaisdell.

Roeser, E. 1957. *Die nichteuklidischen Geometrien und ihre Beziehungen untereinander,* Munich: Oldenbourg.

Shirokov, P. A. 1964. *Sketch of the Fundamentals of Lobachevskian Geometry,* Groningen (the Netherlands): Noordhoff.

Taylor, E. F. and J. A. Wheeler. 1966. *Spacetime Physics,* San Francisco: W. H. Freeman.

Tuller, A. 1967. *Modern Introduction to Geometries,* New York: Van Nostrand Reinhold.

Verriest, G. 1951. *Introduction à la Géométrie Non-Euclidienne par la Méthode Elémentaire,* Paris: Gauthier-Villars.

Wylie, C. R., Jr. 1964. *Foundations of Geometry,* New York: McGraw-Hill.

Mathematically Advanced

Bachmann, F. 1973. *Aufbau der Geometrie aus dem Spiegelungsbegriff*, New York: Springer-Verlag.

Blumenthal, Leonard, and Karl Menger. 1970. *Studies in Geometry*, San Francisco: W. H. Freeman.

Borsuk, K. and W. Szmielew. 1960. *Foundations of Geometry*, Amsterdam: North Holland.

Cartan, E. 1950. *Lecons sur la Géométrie Projective Complexe,* Paris: Gauthier-Villars.

Coxeter, H. S. M. 1968. *Non-Euclidean Geometry,* 5th ed., University of Toronto Press.

Dembowski, P. 1968. *Finite Geometries,* New York: Springer-Verlag.

Ewald, G. 1971. *Geometry: An Introduction,* Belmont (Calif.): Wadsworth.

Henkin, L., P. Suppes, and A. Tarski, eds. 1959. *The Axiomatic Method,* Amsterdam: North Holland.

Hilbert, D. 1971. *Foundations of Geometry,* 2nd ed., La Salle (Illinois): Open Court.

Klein, F. 1968. *Vorlesungen über Nicht-Euklidische Geometrie,* Berlin: Springer-Verlag.

Moise, E. E. 1963. *Elementary Geometry from an Advanced Standpoint,* Reading (Mass.). Addison Wesley

Redei, L. 1968. *Foundation of Euclidean and Non-Euclidean Geometries According to F. Klein,* Elmsford (N.Y.): Pergamon.

Schwerdtfeger, H. 1962. *Geometry of Complex Numbers,* University of Toronto Press.

Sommerville, D. M. Y. 1958. *Elements of Non-Euclidean Geometry,* New York: Dover.

Spivak, M. 1970. *A Comprehensive Introduction to Differential Geometry,* 2 vols., Waltham (Mass.): Publish or Perish Press.

Stevenson, F. W. 1972. *Projective Planes,* San Francisco: W. H. Freeman.

Wolfe, H. E. 1945. *Introduction to Non-Euclidean Geometry,* New York: Holt, Rinehart, Winston.

Axioms

Incidence Axioms

AXIOM I-1.   For every point P and for every point Q not equal to P there exists a unique line *l* that passes through P and Q. (See p. 42.)

AXIOM I-2.   For every line *l* there exist at least two distinct points that are incident with *l*. (See p. 42.)

AXIOM I-3.   There exist three distinct points with the property that no line is incident with all three of them. (See p. 42.)

Betweenness Axioms

To b is between A and C or (ABC)

AXIOM B-1.   If A * B * C, then A, B, and C are three distinct points all lying on the same line, and C * B * A. (See p. 61.)

AXIOM B-2.   Given any two distinct points B and D, there exist points A, C and E lying on $\overleftrightarrow{BD}$ such that A * B * D, B * C * D and B * D * E. (See p. 61.)

AXIOM B-3.   If A, B, and C are three distinct points lying on the same line, then one and only one of the points is between the other two. (See p. 61.)

AXIOM B-4.   For every line *l* and for any three points A, B, and C not lying on *l*:

(i) if A and B are on the same side of l, and B and C are on the same side of l, then A and C are on the same side of l.

(ii) if A and B are on opposite sides of l and B and C are on opposite sides of l, then A and C are on the same side of l. (See p. 64.)

Congruence Axioms

AXIOM C-1. If A and B are distinct points and if A′ is any point, then for each ray r emanating from A′, there is a *unique* point B′ on r such that B′ ≠ A′ and AB ≅ A′B′. (See p. 70.)

AXIOM C-2. If AB ≅ CD and AB ≅ EF, then CD ≅ EF. Moreover, every segment is congruent to itself. (See p. 70.)

AXIOM C-3. If A * B * C, A′ * B′ * C′, AB ≅ A′B′ and BC ≅ B′C′, then AC ≅ A′C′. (See p. 70.)

AXIOM C-4. Given any angle ⧼ BAC (where by definition of "angle" $\overrightarrow{AB}$ is not opposite to $\overrightarrow{AC}$), and given any ray $\overrightarrow{A'B'}$ emanating from a point A′, then there is a unique ray $\overrightarrow{A'C'}$ on a given side of line $\overleftrightarrow{A'B'}$ such that ⧼ B′A′C′ ≅ ⧼ BAC. (See p. 71.)

AXIOM C-5. If ⧼ A ≅ ⧼ B and ⧼ A ≅ ⧼ C, then ⧼ B ≅ ⧼ C. Moreover, every angle is congruent to itself. (See p. 71.)

AXIOM C-6 (SAS). If two sides and the included angle of one triangle are congruent respectively to two sides and the included angle of another triangle, then the two triangles are congruent. (See p. 72.)

Continuity Axioms

ARCHIMEDES' AXIOM. If AB and CD are any segments, then there is a number n such that if segment CD is laid off n times on the ray $\overrightarrow{AB}$ emanating from A, then a point E is reached where $n \cdot CD ≅ AE$ and B is between A and E. (See p. 79.)

DEDEKIND'S AXIOM.   Suppose that the set of all points on a line
l is the union $\Sigma_1 \cup \Sigma_2$ of two nonempty subsets such that no point
of Σ_1 is between two points of Σ_2 and vice versa. Then there is a
unique point O lying on l such that $P_1 * O * P_2$ if and only if $P_1 \varepsilon \Sigma_1$
and $P_2 \varepsilon \Sigma_2$ and $O \neq P_1, P_2$. (See p. 80.)

Parallelism Axioms

HILBERT'S PARALLEL AXIOM FOR EUCLIDEAN GEOMETRY.
For every line l and every point P not lying on l there is at most one
line m through P such that m is parallel to l. (See p. 84.)

EUCLID'S FIFTH POSTULATE.   If two lines are intersected by a
transversal in such a way that the sum of the degree measures of the
two interior angles on one side of the transversal is less than 180°, then
the two lines meet on that side of the transversal. (See p. 107.)

HYPERBOLIC PARALLEL AXIOM.   There exist a line l, a point P
not on l such that at least two distinct lines parallel to l pass
through P. (See p. 150.)

Symbols

AB   segment with endpoints A and B (p. 12)

$\overleftrightarrow{PQ}$ line through P and Q (p. 12)

$\cong$ congruent (p. 13)

$\overrightarrow{AB}$ ray emanating from A through B (p. 14)

$\measuredangle$ A   angle with vertex A (p. 15)

$l\|m$ line l parallel to line m (p. 17)

$l\perp m$ line l perpendicular to line m (p. 23)

$\triangle$ABC   triangle with vertices A, B, C (p. 24)

$\square$ABCD   quadrilateral with successive vertices A, B, C, D (p. 24)

$S \cup T$ union of S and T (p. 25)

$S \cap T$ intersection of S and T (p. 25)

RAA   Reductio-ad-absurdam method of proof (p. 34)

$H \Rightarrow C$ statement H implies statement C (p. 34)

$\sim S$ negation of statement S (p. 37)

$A * B * C$ point B is between points A and C (p. 61)

$AB < CD$ segment AB is smaller than segment CD (p. 75)

$\measuredangle$ ABC $< \measuredangle$ DEF   angle ABC is smaller than angle DEF (p. 77)

$(\measuredangle A)°$ number of degrees in angle A (p. 101)

$\overline{AB}$ length of segment AB (p. 101)

$\triangle$ABC $\sim \triangle$DEF   triangle ABC is similar to triangle DEF (p. 125)

$\Pi(PQ)°$ number of degrees in angle of parallelism associated to PQ (p. 161)

$P(l)$ pole of the chord l (p. 197)

(AB, CD) cross-ratio of ordered tetrad ABCD (p. 205)

$d(AB)$ Poincaré-length of Poincaré segment AB (p. 205)

$d'(AB)$ Klein-length of segment AB (p. 223)

A)(B   open chord with endpoints A and B (p. 187)

Subject Index

Name Index